Even wrong theories can give correct answers for a while.

—Albert Einstein

The End of
Pseudo-Science

The End of Pseudo-Science

Essays Refuting False Scientific Theories Taught in Schools, Colleges, and Universities

Mohammed Abu-Bakr

An American Atomist

iUniverse, Inc.
New York   Bloomington

The End of Pseudo-Science

Essays Refuting False Scientific Theories Taught in Schools, Colleges, and Universities

iUniverse books may be ordered through booksellers or by contacting:

iUniverse
1663 Liberty Drive
Bloomington, IN 47403
www.iuniverse.com
1-800-Authors (1-800-288-4677)

ISBN: 978-0-595-42024-7 (pbk)
ISBN: 978-0-595-86368-6 (ebk)

Printed in the United States of America

iUniverse rev. date: 3/27/2009

Cover Photograph: Variety of metaphysical atoms of the Atomists, as depicted by the Russian-born American science fiction and science writer Isaac Asimov (1920-1992). Courtesy of SCHOLASTIC INC.

Visit author's website at

www.atomism.info

In deep love, respect, and gratitude to my mother, Willie Gray Watson Gardner, for her intellectual interest and enthusiastic support.

Contents

Acknowledgements

I want to thank the following publishers for their permission to print copyrighted material. Any errors or omissions are unintentional.

Dorling Kindersley Limited, for the excerpt from *The Story of Philosophy* by Bryan Magee, copyright 2001.

Hackett Publishing Company, for the excerpts from *Lucretius, On the Nature of Things*, Translated by Martin Ferguson Smith, copyright 2001.

Harper Collins Publishers, for the excerpts from *The Harper Collins Dictionary of Philosophy*, 2nd Edition, by Peter Angeles, copyright 1992.

Penguin Books, for the excerpts from *Lucretius, On the Nature of the Universe*, translated by R. E. Latham, and revised with a new introduction and notes by John Godwin, copyright 1951.

W.W. Norton & Company, for the excerpts from *The Dream of Reason* by Anthony Gottlieb, copyright 2001.

I also want to thank the following people who helped so much in the writing of this book. The reference desk staff at the main Denver Public Library and the Holly Branch for their help, friendliness, and services; the professors of physics at the University of Colorado at Boulder and the many Denver Public Schools science teachers who gave opinions about the theories presented in this book and helped me present the theories as accurately as possible; Mr. Loukas Loukas, a scholar of the Greek language, who helped me spell and understand the key terms used by the Ancient Greek Atomists to expound Atomism; and especially my friends, Wahab Baouchi, a physicist, and Sarah Curry, a math teacher, who have contributed in every way to making this project a reality. Without their vital and enthusiastic support, this project may not have been completed. I am grateful to them for this dedication to me and my project. Always, my thanks goes to my wife, Sandra, who has always supported my research and writing in every way.

Introduction

The End of Pseudo-Science deals with the most serious problem in teaching sci-ence today in our schools, colleges, and universities. This problem is that much of what is taught is not real science. It is pseudo-science. What is pseudo-science? Whatever definitions scientists have given of pseudo-science, I find none so accu-rately descriptive of it as this: that it is science that does not correspond to reality and violates the laws of nature. Reality is that which is real; that which has actual or *objective* existence. It is distinct from the imaginary, that which exists in the imagination only and is, therefore, unreal. It is what things are like in themselves, not what things are like to the observer. Laws of nature are general statements that describe regular sequences of events that are observed to occur in nature with unvarying uniformity under the same condition. For example, we observe that at sea level water boils at 212 degrees Fahrenheit or 100 degrees centigrade, but a change in external pressure raises or lowers its boiling point. Any statement that water does not behave like this would be contrary to reality and a *violation* of a law of nature.

The pseudo-science taught in our schools, colleges and universities is a conse-quence of using textbooks that give instructions in science based on false ideas about the behavior of matter and energy. Matter and energy are the two most basic ideas in modern science. Physicists believe that all of the observable occur-rences in the universe can be explained in terms of the interaction of matter and energy. They claim that matter and energy together make up the physical uni-verse, but they are wrong. Atoms and kenon are the sole constituents of the uni-verse. Physicists define matter as anything that has mass and occupies space. Energy theorists say that energy is the mover of matter. They define energy as the *ability to do work,* such as moving things, changing the shape or the phase of mat-ter, giving heat and light, and making things grow. Notice that the definition of energy as the ability to do work does not tell us what energy is, only what it does or can do. As the American theoretical physicist Richard P. Feynman (1918-1988) explained, "It is very important to realize that in physics today, we have no knowledge of what energy is."[2] Despite this, modern physicists have unwittingly made energy a basic idea in physics and do physics by studying the "supposed" interaction of matter and energy, claiming that they can observe the effects of energy on matter. They offer us various scientific theories that they say explain the behavior of matter and energy. I call them "false energy theories of matter," or simply "energy theories." They include Thomas Young's theory of energy, the big bang theory, the string theory, Albert Einstein's theory of matter and energy, the quantum theory, and quantum mechanics. All of these theories make claims that are contrary to reality and violate laws of nature. They prevent scientists from correctly understanding the laws of nature and developing a deeper understanding of what things are made of and how things work, which is absolutely necessary for the progress of science beyond superficial and ephemeral levels. In other words, all of the energy theories of matter have a pernicious effect on the study of science.

I wrote this book to refute all of the energy theories of matter and to introduce the scientific community and general public to Atomism, also known as the Atomic Theory, as a way to put an end to the teaching of pseudo-science in our schools, colleges, and universities, and to acquaint the scientific community with the proper way to do physics.

Atomism is a theory in natural philosophy of reality and causality. It is the theory people need to know to understand the world we live in.[3] It explains that reality consists of atoms and kenon, and that the interaction of atoms in kenon ultimately causes all the different phenomena in nature. This being true, the proper way to do physics is to study the interaction of the atoms. *If you understand what atoms are, you understand that they must exist.* The Atomists' argument for the existence of atoms is not based on observation, or any form of external evidence, but simply on a particular definition of the meaning of the word "atom." Atoms are the smallest particles of matter that can exist—the ultimate and smallest division of matter. If matter is repeatedly cut up, the end result will be uncuttable pieces of matter or "atoms." The word *atom* comes from the Greek word *ātomos* meaning "uncuttable."[4] Since it is impossible for atoms to come into existence out of nothing or pass away into nothing, they are eternal and indestructible. Atoms are the absolutely solid and unchangeable things from which nature forms, increases, and sustains all things, and into which she again resolves them when they perish. Kenon is the pure empty space that separates atoms and through which atoms move. Without this kenon, it would be impossible for atoms to move. All things are composed of atoms and kenon.[5]

All atoms are in perpetual rapid motion in kenon. Motion is a fundamental property of atoms and it is eternal and indestructible, just as are the atoms themselves. There are four causes of an atom's motion. The first is the eternal *dynamis*[6] (force) of its own *varos* (weight or mass) on kenon which cannot offer resistance to this dynamis and allows the atom to fall straight downward. The second is its *clinamen* or *parenklisis* (swerve) from its straight downward path under the influence of gravitational pull on it by an adjacent and larger atom. Every atom in the universe is attracted to every other atom. The third is its *collisions* with other atoms caused by its *clinamen,* which sometimes generate the patterns of motion by which aggregates of atoms are formed. An aggregate of atoms is a group of atoms gathered together into a whole. Every object in the universe—from the tiny quark to a huge planet—is an aggregate of atoms.

The fourth cause of an atom's motion is its inherent quality of *repellency*. Just as atoms attract each other when they come near each other, they repel each other when they are squeezed together. The atom's dual quality of gravity and repellency is a *mystery*. With every atom possessing the quality of repellency, all aggregates of atoms, in due course of time, go into a *dissipating* motion that destroys their forms and thereby causes them to perish.

Kinema (motion) is a fundamental property of atoms and all kinēseis (changes) in the universe are reducible to the movements of atoms in the kenon. All changes in matter, or aggregates of atoms, are the result of atoms moving in the kenon and combining in different ways. Never forget that atoms are perpetually moving, even within a compound object, but in tight and regular cyclical patterns which make the complex as a whole stable. Atoms are perpetually moving, even within a compound, because the medium through which they move is kenon, which can offer no resistance to their motion.

Since nothing exists but atoms and kenon, and since the medium through which atoms move is kenon, the idea that energy is what makes atoms move is unrealistic and unnecessary. Let it be understood that energy is produced by the motions of atoms; the motions of atoms are not produced by energy. Atoms "carry on" energy whenever they are producing something, such as heat. Energy is not a primary property of the atoms. The primary properties of atoms are size, shape, *varos* (weight or mass), solidity, gravity, position, arrangement, motion, repellency, and *dynamis* (the power or ability, to do, act, or produce a certain effect).

Let it also be understood that energy is an effect produced by the dynamis of moving atoms. Energy theorists confuse energy with dynamis. Atoms have dynamis from their sheer existence, but not energy. Without this dynamis, it would be impossible for atoms to produce energy.

It is important for energy theorists to realize that their energy theory of matter does not and cannot explain what matter is made of and how it changes and moves. The only theory that can do this is Atomism.

Being of Greek origin, Atomism cannot be understood without knowing the meaning of the following key words used by its Greek developers to explain the theory:

aitia, plural *aities*, meaning "cause"
anaphēs, meaning "tangible"
Atomismos, meaning "Atomists' ideas or beliefs about the atoms"
clinamen, meaning "downward swerve"
dynamis, plural *dynameis*, literally meaning "power," "force," or "strength"
energeia, plural *energeis*, literally meaning "activity"
epidrase, plural *epidraeis*, literally meaning "effect"
hylē, meaning the primary substance underlying reality, or the stuff of which something is composed.
kenon, plural *kena*, literally meaning "pure empty space"
kinema, meaning "motion"
kinēsis, plural *kinēseis*, meaning "any kind of change or motion"
parenklsis, meaning "swerve"
varos, meaning weight, or force of mass on kenon

I will use these Greek words in this book whenever I deem it possible and feasible to help me explain what Atomism is, as accurately as possible. I will always use the words *dynamis, kenon, and varos* in an effort to introduce them into the modern technical vocabulary of science.

Many physicists are striving to discover the basic particle of matter—that is, the particle that cannot be divided and is the basic building block of matter. I believe that the real atom described in this book is that particle. Many physicists are also striving to discover a theory that will explain everything. I believe that Atomism is that theory.

I wrote this book to introduce the scientific community and the general public to Atomism. I discovered that none of the science textbooks used in our schools, colleges, and universities introduce students to Atomism. I also discovered that only a few science teachers have heard of Atomism. Interestingly, some science textbooks teach that matter can no longer be thought to be composed of solid atoms, but the atom itself is divisible into many constituent particles—electrons, protons, neutrons, etc. I will show in this book that those textbooks are wrong.

All science textbooks today teach that matter is composed of atoms and call this teaching the "atomic theory," but the particle that they describe as an atom is not an atom. It is a particle mistaken for an atom and misnamed an atom. Consequently, science textbooks distort the true Atomic theory.

Before writing this book, I asked myself, "What would be the best way to introduce both the scientific community and the general pubic to Atomism?" I decided to write essays dealing with such scientific topics as atoms, matter, motion, gravity, energy, light, and time. All of the essays are meant to be stimulating and effective refutations of false scientific theories about these things. Each essay contains some information that exists in the others. I considered that information to be "necessary redundancy"[7] for the purpose of showing when a theory of a thing is incompatible with atomic reality and atomic causes.

I frequently discussed my essays, especially those refuting theories held by Einstein, with others. I was asked on several occasions by individuals, scientists included, who hold Einstein to be their intellectual icon: "Do you think you are smarter than Einstein?" I do not proclaim to be smarter than Einstein or anyone else, but I do wonder why it should be inconceivable that another human being could be as smart or smarter than Einstein. I only know that in my intellectual and scientific judgment, his theories about matter, energy, atoms, light, gravity, space, and time are wrong.

Except for the first two essays, none of the essays in this book required a lengthy explanation of the scientific theories that they are designed to refute. I wrote all of them with the understanding that if the basic principles of a given theory are false, the rest of the theory will be false. To me, it would have been of little value to my effort to refute the theories by examining secondary and tertiary principles. If theories are not rejected and abandoned when their basic principles are proved false, they are not likely to be rejected and abandoned on the basis of other false principles derived from them.

All of the essays contained in this book are based on the science of reasoning, logic, and the "principle of consistency," which holds that "truth does not involve contradiction." They are also based on the belief that hard empirical evidence supersedes inferred mathematical conclusions. As you read them, be cognizant of the fact that differences in definitions of ideas, things, activities, and effects can lead to great differences in conclusions. If the definition of a thing does not fit reality, the definition is false and will lead to a false conclusion. If the description of a thing does not fit the definition of a thing, it cannot be that thing.

Essay #1, *Why Scientists Don't Know What Atomism Is*, explains what Atomism is and why it died out in modern science.

Essay #2, *It's a Thom, Not an Atom*, explains what an atom is and distinguishes it from the "thom," the particle mistakenly identified as an atom in modern physics.

Essay #3, *A Refutation of the Big Bang Theory*, explains what is meant by the word "universe," why the universe is eternal and could not have been created by an explosion of a point-like object.

Essay #4, *A Refutation of the String Theory*, explains why strings could not possibly be the basic building blocks of the universe.

Essay #5, *Why Physicists Don't Know What Energy Is*, explains that the word energy is a misnomer for the fundamental property of matter called *dynamis*, which really means the ability of a thing to do, act, or produce an effect. It shows that the idea that energy is the ability to do work is misleading.

Essay #6, *What Makes Matter Change and Move?*, explains that motion is eternal and a fundamental property of the self-moving atoms of which the universe and all things in it are composed.

Essay #7, *Einstein Was Wrong: E≠mc²*, explains that matter and energy are not equivalent, and that matter cannot be changed into energy and energy into matter.

Essay #8, *A Refutation of Einstein's Description of Light*, explains that light is composed of tiny massy atoms, and not particles without mass called photons, as claimed by Einstein.

Essay #9, *A Refutation of Einstein's Theory of Time*, explains what time is and that time cannot be slowed down by the near light-speed of an object, as claimed by Einstein.

Essay #10, *A Refutation of Einstein's Theory of Gravitation and Curvature of Space*, explains that gravity does not exist as a property of empty space between objects and that it cannot change the form and structure of space, as claimed by Einstein.

Essay #11, *Einstein's False Conception of the Universe*, explains how Einstein's conception of the universe is deeply flawed with shallow reasoning and faulty logic that led him to falsely restructure the laws of physics.

Essay #12, *Why Quantum Theory and Quantum Mechanics are Wrong*, explains that there is no such thing as the study of matter and energy at the subatomic level and refutes quantum theory's claim that matter is not made up of solid atoms. It also shows that there is no such thing as the study of energy levels in an atom, as claimed by quantum mechanists.

Essay #13, *Biologists' False Claim About the Theory of Spontaneous Generation,* explains how the Atomic theory of spontaneous generation of life from nonliving matter is ultimately true.

Essay #14, *Myths About Black Holes and Baby Universes,* explains that black holes are not pathways to baby universes, but only atomizers of matter.

When you have finished reading the essays in this book, you will understand that Atomism is a theory of reality and causality that provides a view of our Atomic world that is necessary to every well informed person.

Mohammed Abu-Bakr
December 2006
Denver, Colorado

Essay #1
Why Physicists Don't Know What Atomism Is

Physicists are scientists who study what things are made of and how they work. *Atomism* is a theory that the universe and all of the things in it are made up of *atoms* and *kenon*, and that the interaction of these atoms in kenon ultimately causes all the different phenomena in nature. Atomism explains that the ever-shifting arrangement of atoms in kenon is the ultimate reality and cause behind all appearances.

The existence of atoms was discovered about 2,400 years ago by an ancient Greek philosopher named Leucippus, who studied the nature of reality. He realized that the things that appear to us as finite bodies of matter (rocks, trees, etc.) were made up of smaller parts, and that no finite body of matter could be made up of an infinite number of parts. He thought that if a finite body of matter (hylē), anything *anaphēs* (tangible), was repeatedly cut up, the end result would be uncuttable pieces of matter or "atoms." The word *atom* comes from the Greek word *ātomos*, meaning "uncuttable." Leucippus maintained that atoms are the smallest particles of matter that exist – the ultimate and smallest division of matter.

Leucippus accepted the Greek notion of kenon ("empty space"). He held kenon to be the pure empty space that separates atoms and in which they move. He explained that reality consists of both atoms and kenon, and that atoms and kenon are equally real because they both make up the universe and all of the things in it. He was credited with having evolved the Atomic theory of matter.

Leucippus' illustrious student named Democritus defended and continued to develop Leucippus' theory. He maintained that nothing exists but atoms and kenon. He ascribed size, shape, and solidity, or impenetrability to the atoms. He also combined the theory of the atoms with the theory of elements – fire, air, earth, and water. He said that these elements arise from the various arrangements and positions of the atoms.

I, myself, ascribe the following fundamental properties to the atoms: size, shape, solidity, *varos* (weight, force of mass), gravity, arrangement, position, *kinema* (motion), repellency, and dynamis. I describe these properties of the atoms in the following way:

Size of Mass: Being the smallest particle of matter that exists, an atom is infinitesimally small. It is so small that it cannot even be seen with the most powerful electron microscope, which can magnify an object up to a million times. We can only estimate the size of an atom based on the size of the smallest known aggregate of atoms called the "U quark," which has a mass approximately 1×10^{-27} grams. Since the U quark is made up of atoms, the mass of an atom is at least less than 1×10^{-27} grams, perhaps by a factor of 1,000. Therefore, I estimate the mass of an atom to range from approximately 1×10^{-30} grams to as close to zero as one pleases.

Shape: Since atoms are too small to be seen, we do not have any way of knowing what the shapes of the atoms are. Obviously, the great variety of forms of matter produced by atoms shows that atoms are probably not all the same shapes. Atoms of oil could be round, atoms of a rock might have an irregular shape, and so forth.

Solidity: Every atom exists as a solid thing, with a definite size, shape, and well-defined and impenetrable surface that separates it from all other atoms.

Varos: This is the force with which an atom presses downward. It is also a measure of the pull of gravitation on an atom. Hence, it is the weight of an atom.

Gravity: This is the force that atoms and objects composed of atoms exert on each other because of their mass. Each atom attracts every other atom in the universe.

Arrangement: Every object is a particular geometric arrangement of fixed numbers of atoms.

Position: Every atom has a place where it is located within an object.

Motion: All atoms are in perpetual motion. They possess a certain original momentum that can be transferred to other atoms by impact. All changes in matter are the result of atoms moving and combining in different ways.

Repellency: Just as every atom in the universe attracts every other atom, every atom in the universe also repels every other atom, preventing actual contact. The force of repellency gets stronger the closer atoms come together. This is the force that keeps atoms separated by kenon, however small, and causes every object in the universe to have a porous texture and to eventually disintegrate.

Dynamis: This is the ability or capacity of atoms to do, act, or produce a certain effect. Every atom has dynamis from its sheer existence. The dynamis of an atom is equivalent to its mass in motion.

Atoms do not possess properties such as color, taste, smell, sound, or being hot or cold. These properties are produced by the *energeia* (activity) of the atoms upon sense organs.

I will now acquaint you with basic principles of Atomism. A basic principle is a general truth on which other truths are based. Atomism says:

- *Nothing can be created out of nothing.* This is a *self-evident truth* for anyone who has the *intelligence* to understand that a thing that does not exist cannot be created out of something that also does not exist. This is the most basic and necessary self-evident truth upon which Atomism is built as a logical system of undeniable reality.

- *Nothing can be destroyed into nothing.* This principle is based on the observation that nature never totally destroys anything, but only changes the form of a thing, or breaks it up into constituent parts.

- *Matter exists in the form of invisible particles.* This principle is based on sensory experiences of things we never see, such as the wind, sound, odors, cold, and heat; and yet all of these things must be of a material nature since they have the dynamis (power) to set upon our sensory organs. For nothing can touch or be touched unless it is material. Certainly these things that bring us the sense of wind, sound, odors, cold, and heat are invisible particles of matter.

- *Matter cannot be created or destroyed into nothing, but only broken up into constituent atoms.* This principle is based on the observation that matter can be repeatedly cut up into smaller and smaller pieces and the logic that the end result of cutting up matter will be uncuttable pieces of matter or "atoms," which are eternal and indestructible particles.

- *Atoms and kenon (empty space) are the only ultimate realities.* This principle is based on the fact that atoms are the ultimate and smallest division of matter and observation that matter is constantly in *motion*, which requires a kenon. Nothing else exists but the *imaginary*.

- *All other things are properties or accidents of atoms and kenon.* This principle is based on the fact that all matter is an aggregate of atoms and that the properties and behavior of matter arise from these atoms, which move randomly in kenon.

- *All matter is made up of atoms and kenon.* This principle is based on the fact that all matter is an aggregate of atoms and the observation that all matter is *porous*.

- *The atoms are absolutely solid, simple, and everlasting.* This principle is based on the fact that atoms are the smallest particles of matter that can exist and are, therefore, composed of a single type substance that does not have qualities of vacuity and destructibility.

- *Being absolutely solid and simple, the atoms are indivisible.* This principle is based on the observation that the harder an object is, the more difficult it is to cut or divide into parts. This is because the harder an object is, the more tightly compact the aggregate of atoms of which it is composed. The more tightly compact the aggregate of atoms composing an object, the smaller or narrower the interstices between the atoms. To divide an aggregate of atoms into parts, we must pierce the interstices with a sharp-edged instrument or knock it into parts with a sharp blow. Since the individual atoms that make up an object are absolutely solid without interstices, they cannot be divided into parts.

- *The atoms cannot undergo change.* This principle is based on the observation that the harder an object is, the more resistant it is to change. Based on this fact, it is reasonable to conclude that the atoms, being eternal, absolutely solid, and indestructible, retain their properties under all conditions.

- *Although physically indivisible, the atoms have parts, which are the minima of extension and magnitude.* This principle is based on the observation that all objects are three-dimensional, having length, width, and height. Since atoms are the building blocks of these objects, they themselves must be three-dimensional objects. All three-dimensional objects have size or extent, no matter how infinitesimally small they may be. If one could see an atom, one would be able to point to the part of the atom nearest the top (or highest or uppermost part, side or surface) of an atom, or to the middle (or the part, point, position, or area in the center, equally distant from the ends, sides, or edges) of an atom, or to the bottom (or the lowest part, or the underside or under surface) of an atom.

I think of these basic principles of Atomism in the same way that the British philosopher and mathematician Bertrand Russell (1872-1970) thought of mathematics. They "rightly viewed, possess not only truth, but supreme beauty—a beauty cold and austere, like that of sculpture.[8]

Being the chief type of naturalism, Atomism holds that everything comes from nature and there is nothing beyond or above nature. It rejects the supernatural and explains that all things are subject solely to the laws of nature.

Being the chief type of materialism, Atomism holds the following basic principles of materialism to be self-evident truths:

- All things are basically material.
- Nothing can exist which is non-material, except pure empty space.
- Space is either absolutely filled or absolutely empty.
- No two things can occupy the same space at the same time.
- A thing cannot exist in two different places at the same time.

Atomism is the only theory that can correctly explain our Atomic world. Unfortunately, it was eclipsed by Christian Scholasticism in the 1000's and has not been revived as a theory of reality and causality, only as a theory of matter in 1649, by the French philosopher Pierre Gassendi. In 1897, the English physicist Joseph J. Thomson claimed to have discovered particles that were smaller than atoms called "corpuscles." Later these particles became known as "electrons." Thomson developed a model of this idea of the atom, showing electrons to be subatomic particles. Thomson's idea that electrons are smaller than atoms was wrong and hence so was his model of the atom. But his idea that electrons are smaller than atoms, and his model of the atom showing electrons to be subatomic particles, were accepted by the scientific community as true without any proof. Interest in the true atom of the Atomists and Atomism died out.

Because of this, physicists today do not know what Atomism is. The word "Atomism" is not even generally familiar to them. They do not know that they need to know Atomism to understand the Atomic world we live in.

Essay #2
It's a Thom, Not an Atom

Let me begin this discussion by explaining that I am an Atomist, a believer in Atomism, the natural philosophy that originated the idea of the atom. I know that modern scientists, after nearly 255 years, still mistake the wrong particle of matter for the atom. I want to correct this mistake, primarily to defend Atomism's correct idea of the atom, which is essential to explaining the "true" nature of the universe.

I propose that the particle of matter that modem scientists wrongly call an atom be called a *thom* (tom), for reasons I will explain later. As I tried to think of the best way to convince present-day scientists that they mistake the thom for the atom, I thought of the following argument once made by the great Greek philosopher and scientist, Aristotle (384-322 B.C.): "He who studies how things originated and came into being ... will achieve the clearest view of them."[9] I agree with this very insightful statement made by Aristotle. For this reason, I will try to carry you through a careful study of the historical development of Atomism, also called the "Atomic Theory of Reality and Causality" which will clearly prove that modern scientists mistake the thom for the atom, and that this mistake is one of the most serious blunders of modem science.

Atomism holds that the ultimate, unchangeable reality consists of atoms in motion and kenon. This philosophic theory was developed by the early Greek philosophers Leucippus (fl. c 450 B.C.), Democritus (c.460–.370 B.C.), and Epicurus (341–270 B.C.), and the Roman philosopher and scientific poet Lucretius (c.98–55 B.C.).

Leucippus studied the nature of reality and came to believe that matter is the ultimate substance underlying reality. He had the logical idea that if matter were repeatedly cut up into finer and finer pieces, eventually a point would be reached where only the smallest uncuttable pieces of matter would be left. He held that these smallest pieces of matter would be uncuttable, not just because they would be too small to be cut, but because they would be absolutely solid. Leucippus' illustrious student and follower Democritus called these smallest uncuttable pieces of matter atoma, which means "uncuttables." Atoma is the plural of ātomos. The word *atomos* is made of the privative letter *a*, not, plus *topos*, able to be cut, and means exactly "uncuttable." Our English word *atom* comes from the Greek word *atomos*. Leucippus and Democritus used the word *atomos* primarily to designate and describe the ultimate, unchangeable reality underlying changing forms of matter (hylē). They added the suffix *ismos* to coin a noun that names Atomists' ideas or beliefs about the atom.

Lucretius explained that atoms are the things from which nature forms, increases, and sustains all things, and into which she again resolves them when they perish. He gave them such names as "matter," "raw materials," "generative bodies," "seeds of things," "first bodies," and "primary particles," because they come first and everything is composed of them. Lucretius considered the atoms to be nature's ultimate instruments of work.

As the Atomists explain, the atom is the smallest particle of matter that can exist—the ultimate and smallest division of matter. Atoms are the ultimate unchanging building-blocks of all matter. Hence, the properties and behavior of matter arise from the properties and behavior of the atoms of which it is composed. Nothing can come into existence without atoms. Being the ultimate building-blocks of matter, atoms have such properties as size, shape, weight, solidity, gravity, position, arrangement, repellency, movement, and dynamis. The Harper Collins Dictionary of Philosophy, 2nd Edition, explains that: "Atoms are eternally present, simple, separate, irreducible to anything further, impenetrable, unchangeable (their essential nature remains eternally the same), and invisible to the naked eye. There are infinitely many atoms in existence. Atoms in themselves do not possess qualities such as color, taste, heat, or smell. These qualities are produced by the activity of the atoms upon sense organs. The atoms are eternally in motion in empty space, colliding with each other and forming objects. Their sizes and geometric shapes interlock into configurations producing the variety of existing things. The existence of any individual thing can be explained in terms of the arrangements of the positions and figures of the atoms of which it is composed.[10]"

According to the Atomists, the three most important characteristics of the atoms are their absolute solidity, indivisibility, and eternal self-movement. They are infinite in number and vary in size, weight, shape, and arrangement, but not in substance. Atoms have come together by chance to form our world and the things in it as well as innumerable other worlds.

No one knows for sure what Leucippus and Democritus' visualization of an atom was, but modem Atomists say that the easiest way to picture an atom is to think of it as being similar to a "solid sphere," or "ball bearing," that cannot be cut or divided into parts. According to Democritus, atoms might differ in size, weight, and shape, but they are all made of the same kind of "stuff." The "solid sphere," or "ball bearing" model of the atom, is a good model. It fits together the absolute solidity and indivisibility of atoms. Democritus explained that knowledge of atoms comes from the intellect, not from the senses. This means that the existence of atoms is not amenable to scientific check. However, scientists must accept the existence of atoms on the basis of their belief in the impossibility of infinite divisibility of matter.

As the English classicist John Godwin explains: "The indivisible nature of atoms is 'proved' by the continued existence of matter-if there were not some level beyond which matter could not be divided any further, then everything would have dissolved into nothing long ago."[11]

Another English classicist named Martin Ferguson Smith gives us a more detailed argument in support of the indivisible nature of atoms. He says: In what form does compounded matter exist? It exists in the form of an infinite number of absolutely solid, indivisible, unmodifiable, imperishable particles or atoms. The Greek word *atomos* means "indivisible," and so the modern atom, which has been split into subatomic particles, is really misnamed.... [U]nless the elements of matter were indivisible, and indestructible, there could be no permanence for the universe, and indeed no existence, for everything would long ago have been reduced to nothing.[12]

The Atomists say it is ridiculous for anyone to believe in the infinite divisibility of matter, and that it makes perfectly good sense to believe that matter can only be broken up into constituent atoms. I believe that anyone who believes in the infinite divisibility of matter has a poor sense of logic. Matter exists in the form of finite objects, things having limits or boundaries. It is impossible for things that have the quality of being finite to be infinitely divisible.

This brings us to a discussion of the aspect of the Atomic Theory, which is unjustifiably neglected by modem science: the kenon (void). *The Harper Collins Dictionary of Philosophy*, 2nd Edition explains: "The atomists accepted the notion of a void [kenon], an empty space or vacuum and called it a not—being (a nothing) as opposed to a being (the self-moving eternal, material atoms). The void possesses no qualities whatever, no powers, no potentiality, no [tangibility] in any way. It is regarded as pure empty space, in which or at which there is absolutely nothing present. The void is the place that atoms occupied before they moved to another place and the place that would be occupied shortly by other atoms. The atomists defended the void with arguments such as the following:

Objects and the atoms of which they are composed could not move unless a void existed between the atoms. That a void exists in things is evidenced in that some objects can be compressed and some absorb liquids. In compression the atoms are pressed into the void existing between them. In absorption the atoms of a liquid enter into the empty interstices between the atoms of the object and occupy that void. It also is in reference to a void that things can be distinguished, separated, and classified."[13]

Many people reject the notion of the void. They believe that a subtle matter called "ether" fills the empty space that Atomists say exists. They say that all space is occupied by something and that no space can be unoccupied. To me, this belief is totally wrong. I observe that things are in motion, moving out of one place into another. Since no two things can occupy the same space at the same time, they must be moving into existing empty space.

Lucretius explained the existence of the void and its necessity for the functioning of things, saying:

"There is vacuity in things ..., by vacuity I mean intangible and empty space. If it did not exist, things could not move at all. For the distinctive action of matter, which is counteraction and obstruction, would be in force always and everywhere. Nothing could move forward, because nothing could give it a starting point by receding. As it is, we see with our eyes at sea and on land and high up in the sky that all sorts of things in all sorts of ways are on the move. If there were no empty space, these things would be denied the power of restless movement, or rather, they could not possibly have come into existence, embedded as they would have been in motionless matter.

"Besides, there are clear indications that things that pass for solid are in fact porous. Even in rocky caves a trickle of water seeps through, and every surface weeps with brimming drops. Food percolates to every part of an animal's body. Trees grow and pour forth their fruit in season because their food is distributed throughout their length from the tips of the roots through the trunk and along every branch. Noises pass through walls and fly into closed buildings. Freezing cold penetrates to the bones. If there were no vacancies through which the various bodies could make their way, none of the phenomena would be possible.... If two broad bodies suddenly spring apart from contact, all the intervening space must be void until it is occupied by air. However quickly the air rushes in all round, the entire space cannot be filled instantaneously. The air must occupy one spot after another until it has taken possession of the whole space. If anyone supposes that this consequence of such springing apart is made possible by the condensation of air, he is mistaken. For condensation implies that something that was full becomes empty, or vice versa. And I contend that air could not condense so as to produce this effect; or, at any rate, if there were no vacuum, it could not thus shrink into itself and draw its parts together.[14]

"If there were not empty space, everything would be one solid mass; if there were no material objects with the property of filling the space they occupy, all existing space would be utterly void. It is clear, then, that there is an alternation of matter and vacuity, mutually distant, since the whole is neither completely full nor completely empty. There are therefore solid bodies, causing the distinction between empty space and full. And these, as I have just shown, can be neither decomposed by blows from without nor invaded and unknit from within nor destroyed by any other form of assault.

"For it seems that a thing without vacuum can be neither knocked to bits nor snapped nor chopped in two by cutting; nor can it let in moisture or seeping cold or piecing fire, the universal agents of destruction. The more vacuum a thing contains within it, the more readily it yields to these assailants. Hence, if the units of matter are solid and without vacuity, as I have shown, they must be everlasting.

"Yet again, if matter in things had not been everlasting by now they would have gone back to nothing, and that things we see would be the product of rebirth out of nothing. But, since I have already shown that nothing can be created out of nothing nor any existing thing be summoned back to nothing, the atoms must be made of imperishable stuff into which everything can be resolved in the end, so that there may be a stock of matter for building the world anew. The atoms, therefore, are absolutely solid and unalloyed. In no other way could they have survived throughout infinite time to keep the world renewed."[15]

As you can see, Lucretius upheld the Atomic Theory, which holds that all matter is made up of atoms as well as kenon (nothingness, pure empty space that exists between atoms and in which atoms move). The early Atomists held that atoms were not in contact, but that kenon existed between them, claiming that otherwise motion would be impossible. In modern science, the definition given to us of the Atomic Theory only indicates and emphasizes that all matter is made up of atoms.

Leucippus held that "what is [that is, the atoms] is no more real than what is not [that is, the kenon]"[16] He understood that atoms and the kenon are both equally real because they both make up the universe. Understanding this fact, he held that reality consists of both the atoms and the kenon.

Democritus held that nothing exists but atoms and the void. Epicurus accepted the Atomic Theory of Democritus. As the English philosopher Bryan

Magee explains: "Epicurus believed that all there was in the material universe were atoms and empty space, nothing else. Since it is impossible for atoms to come into existence out of nothing or pass away into nothing they are eternal and indestructible. However, their movements are unpredictable, and no combination that they form ever endures. For this reason, physical objects, all of which are combinations of atoms, are ephemeral. Their life is always a story of atoms coming together and then, eventually, dispersing again. All change in the universe consists either of this endlessly repeated process or of the objects thus formed in space."[17]

Atomists maintain that the interaction of atoms ultimately causes all the different phenomena in the universe.

Lucretius said that atoms make up objects just as letters make up words. As I explained earlier, he called atoms 'first bodies,' because they come before anything else. He said that: "Everything is composed of small "first bodies, tiny particles made up of a few 'minima' or 'least parts' which cannot be separated. These 'first bodies,' atoms, are solid, indestructible, and of infinite number. They are mixed with void to make objects of greater hardness or softness, strength or weakness."[18]

Lucretius explained that: "Although every atom is minute, it has a certain magnitude and therefore, despite being physically indivisible, is mentally divisible into a limited number of "smallest parts," that is to say, parts that are the minima of extension and magnitude."[19]

Lucretius made it perfectly clear that atoms are not things that are merely geometric points without any size at all, but rather minute masses with magnitude. If one could see an atom, one would be able to point to the part of the atom nearest the top (or highest or uppermost part, side or surface) of an atom, or to the middle (or the part, point, position, or area in the center, equally distant from the ends, sides, or edges) of an atom, or to the bottom (or the lowest part, or the underside or under surface) of an atom. All atoms are three-dimensional objects.

After Lucretius, for nearly 2,000 years, the theory of atoms was largely ignored. It was largely ignored during the Middle Ages, the period of European history between ancient and modern times, from about 500 A.D. to 1450 A.D., partly because it was rejected by Aristotle who acknowledged no limit to the splitting of things, no rest from crumbling, no smallest unit of matter, although everyone could see that every object has an ultimate point that seems to our senses to be the smallest, from which we may infer that the things we cannot perceive also have an ultimate point which actually is the smallest. Aristotle's theories dominated medieval philosophy and science. The theory of atoms was also ignored because of the dominance of the church and its concern with spirituality. The intellectual movement of the Middles Ages was Scholasticism, an attempt to buttress Christian faith with formal reasoning. The theory of atoms, or Atomism, had to be ignored because it was incompatible with Scholasticism, which held God to be the ultimate reality and creator of the universe.

As the English philosopher Anthony Gottlieb tells us: "Early Christian thinkers seized every opportunity to condemn atomism and to discourage the study of it. Their hostility had nothing to do with the fact that atomism says that all matter consists of indivisible particles, a subject on which neither Jesus nor the Old Testament prophets ever expressed an opinion. Early Christians abhorred atomism for two main reasons. First because it tried to explain everything in terms of mechanisms, not the activities of a deity whom it renders redundant. Secondly because it held that there is no life after death, since all things-including souls and, if there happen to be any lurking in the background, gods-are purely temporary configurations of atoms that will eventually dissolve back into chaos."[20]

Although the theory of atoms was largely ignored for nearly 2000 years, it survived among some scientists. In 1649, the French philosopher and scientist Pierre Gassendi (1592-1655) revived the theory. He taught that matter was composed of many kinds of atoms. He said that these atoms differed in size and shape and that they moved about in space. He thought that all the properties of matter could be explained by the sizes and shapes of atoms. His theory of the atoms was very much like Democritus' theory. During his time, the theory of the atoms began to take hold in the minds of scientists.

In 1661, the Irish scientist Robert Boyle (1627-1691) published a famous book, *The Skeptical Chymst*, in which he stated his belief that matter was made up of atoms. He believed that the primary qualities of atoms were size, shape, order, texture, solidity and motion.

Other scientists who were the founders of modem science such as Francis Bacon (1561-1626) and Isaac Newton (1642-1727) of England, and Galileo Galilei (1564-1642) of Italy believed in atoms. But those scientists could add little more to the theory of atoms than Leucippus, Democritus, Epicurus, and Lucretius had described. These philosophers and scientists had succeeded in providing future philosophers and scientists with the foundational tenets of atomic physics and atomic chemistry. They made it clearly understood that it is the interaction of tiny atoms that ultimately causes all the different phenomena in the universe, and that all matter is made up of "uncuttable" atoms.

In 1750, Ruggero or Rudjero Giuseppe Boscovich (1711-1789), an astronomer and mathematician born in what is now Croatia, suggested that Leucippus and Democritus might have been wrong in believing that atoms are "uncuttable." He invented a new theory that atoms contain smaller parts, and so forth down to the basic building blocks of matter. He thought that these building blocks must be geometric points with no size at all. Boscovich's new theory of the atom was definitely wrong. As I explained earlier, Leucippus and Democritus' absolutely solid, uncuttable atom is the smallest particle of matter that can exist—the ultimate and smallest division of matter. Atoms themselves are the ultimate building blocks of all matter. They are irreducible to anything further. Another proof that Boscovich's new theory of the atom was wrong is that he believed that geometric points with no size at all could be the basic building blocks of matter. An aggregate of geometric points with no size, or magnitude, at all would add up to nothing. Although it is easy to see that Boscovich's new theory of the atoms was completely wrong, incredibly, most atomic physicists as well as most atomic chemists of today accept a modem form of this theory.

Boscovich, being a mathematician, based his new theory of the atom on the fact that an atom has magnitude and that a magnitude can be mathematically subdivided. The fact that a magnitude can be mathematically subdivided does not change the fact that an absolutely solid atom cannot be physically divided. Not even mathematics can reduce matter to nothing or create matter out of nothing. There must be a smallest particle of matter that can exist; that particle of matter is and must be the "uncuttable" or "indivisible" atom.

During the late 1799's chemists discovered that they could combine elements to form compounds only in certain fixed proportions according to mass. In 1808, an English chemist named John Dalton (1766-1844), who evidently did not accept Boscovich's new theory of the atom, tried to revive Leucippus and Democritus' theory to explain this discovery. He rephrased their theory to make it easier for other chemists to understand. He said:

"1. All elements are made up of atoms. Dalton believed that no element is infinitely divisible; that there is an ultimate particle of every element or substance, the uncuttable or indivisible atom.

"2. All atoms of a given element are identical. This idea explains why an element always has the same properties.

"3. The atoms of a given element are different from those of any other element. Dalton believed that the atoms of different elements differ in size, shape, and weight, but not in substance.

"4. Atoms of one element can combine with atoms of other elements to form compounds. A given compound always has the same relative numbers and types of atoms. The first part of this statement is certainly true since the atoms of one element are the same kind of "stuff," or substance, as the atoms of all other elements. As for the second part of the statement, modem scientific experiments show that in every compound, the elements are combined with each other in a definite ratio by weight. Hydrogen and oxygen, for example, combine in a ratio of 1 to 8 to make water. Iron and sulfur combine in a ratio of 7 to 4 to make iron sulfide. A compound forms when iron and sulfur are heated together. In the final analysis, all compounds are made up of atoms.

"5. Atoms are indivisible in chemical processes. That is, atoms are not created or destroyed in chemical reactions. A chemical reaction simply changes the way the atoms are grouped together. Dalton imagined atoms to be similar to tiny ball bearings or marbles (solid spheres) that are impossible to break or be cut or be divided. He maintained that atoms cannot be cut, or divided, but always act as wholes. Modem science agrees that atoms do not divide to form compounds. However, compounds of atoms can be divided. Only two kinds of material objects exist, individual atoms and compounds of atoms."

Dalton also held that atoms are immutable. That is to say, they cannot undergo change. He presented modern chemists with a flawless atomic theory that could explain how elements made up of "uncuttable" atoms can be combined to form compounds.

However, in 1897, an English physicist named Joseph John Thomson (1856-1940), claimed that he discovered that atoms are "cuttable," and that he made this discovery while studying the rays that travel between charged metal plates in a vacuum tube. He claimed that he discovered particles that were smaller than atoms in cathode rays, which was certainly not possible. These rays were seen passing between high voltage terminals in a glass tube filled with low pressure gas. The particles had a negative electric charge. Thomson had discovered the electron. He received the Nobel Prize in physics for that discovery in 1906. Before this, in 1904, he constructed a model showing an atom split in half, representing a positively charged sphere, embedded with electron-like seeds in a watermelon. Electrons are larger than atoms, and they themselves, like everything else, are made up of atoms. In other words, electrons are compounds of atoms, or atomic compounds. Even if scientists discover smaller particles than electrons, they can only be compounds of atoms, if they are not atoms themselves.

Evidently, Thomson was not aware of the fact that the ancient Atomists showed through sound logic that an atom is an absolutely solid, uncuttable particle without an internal structure and has a surface that cannot be penetrated. Thomson's model was definitely not a correct representation of an atom. I propose that his model, and any subsequent modified model based on his, be called a "thom". I propose this for several very important reasons. First of all, it will give Thomson's model a suitable name and correct any mistaken idea that his model was a true representation of an atom. Second, it will make a distinction between the thom and the atom. Third, it will make scientists realize that if their description of a particle does not fit the definition of an atom, which means "uncuttable," it cannot be an atom. An uncuttable atom can only be an uncuttable atom, nothing else.

Although Thomson's model of the atom was wrong, other scientists did not know this. They assumed that he was right in considering the electron to be part of the atom. Consequently, in error, they accepted his conceived particle *as an* atom. Some even proposed other erroneous models of the atom.

In 1911, the English physicist Ernest Rutherford (1871-1937), a former student of Thomson's, theorized the existence of an atomic nucleus, which was actually a theory of the existence of a thomic nucleus. He also stated that the nucleus is surrounded by electrons traveling at tremendous speeds through the thom's outer regions. He suggested that the nucleus contained positively charged particles called "protons" (Greek for "first things"). He demonstrated their existence in 1919 by bombarding thin sheets of gold with alpha particles. Most of the particles passed through the sheets, which showed that the gold thoms must consist chiefly of empty space. But some particles bounced back as if they had hit something solid. Rutherford concluded that since he found spaces in the gold sheets that atoms may not be the solid spheres Dalton said they were. This was an incorrect conclusion, for two reasons. First of all, the gold sheet was not an atom and thus it was not an atom that he bombarded with alpha particles. Second, the gold sheet was a compound of atoms with vacuity, empty spaces between each atom. It was only natural that some particles passed through the empty spaces of the sheets, while other bounced back after hitting solid atoms, which proved that atoms contain no empty spaces. Rutherford thought that the particles had been reflected by a strong force from a tiny but heavy nucleus of an atom, which no solid atom has. It was a thom that he imagined to have a nucleus, not an atom.

Rutherford constructed a model of the thom's mass concentrated in a positively charged nucleus surrounded by electrons. He reached the conclusion that the thom resembled a miniature solar system,[21] where "planets" were the electrons and the "sun" was the nucleus. His theory of the thom did not explain the arrangement of electrons in thoms. In 1913, however, a description of the electron structure was proposed by Neils Bohr (1885-1962), a student of Rutherford.

Neils Bohr suggested that electrons could travel only in a certain set of orbits around the nucleus and were only "allowed" certain energies. He found that electrons with the lowest allowed energy orbit closest to the nucleus, and electrons with the highest allowed energy orbit furthest away. Electrons in a thom behave as if they are stacked, lowest energies first, in "shells" around the nucleus. An electron can jump to a higher level by absorbing energy.

In 1932, James Chadwick (1891-1974), an English physicist who was, like Bohr, a student of Rutherford, discovered another particle in the nucleus-the neutron, an uncharged particle of about the same mass as the proton. Chadwick observed that beryllium, when exposed, released an unknown radiation that in turn ejected protons from the nuclei of various substances. Chadwick interpreted this radiation as being composed of particles of mass approximately equal to the proton, but without electrical charge. He called these particles "neutrons" (Greek for "swimming").

Reflecting back to the discovery of the nucleus in 1911, I think of the fact that bombarding certain thoms with particles from radioactive materials could disintegrate their nuclei, releasing energy (atoms in motion). The heaviest nuclei, those of uranium, can be split by neutrons in this way. Otto Hahn (1879-1968) and Lise Meitner (1878-1968) discovered that the uranium nucleus splits in half, or "fissions," and gives out further neutrons (tiny atomic compounds).

Let me stress this fact: It was the thom that was split, not the "uncuttable," or "unsplittable," atom. In discussing this fact, Gottlieb explains that: "What we now call 'atoms' have famously been split and are no longer regarded as the basic building blocks of matter. In their place we now have our even smaller elementary particles, which are divided into the two classes of quarks and leptons. But this is a red herring. It does not mean that the central thesis of ancient atomism turned out to be false. The fact what we call 'atoms' turned out after all not to be the fundamental constituents of matter shows only that modern physics were too hasty in dignifying these particular particles with the name of 'atoms'. Nineteenth-century scientist called the particles that they investigated 'atoms' because they believed falsely, as it turned out, that these particles could not be split. The atomism of Democritus and Leucippus will not be refuted until it is shown that there are no unsplittable basic particles, whatever you wish to call them."[22]

In the 1960's scientists discovered that protons and neutrons are composed of even smaller particles. In 1964 the American physicist Murray Gell-Mann named them quarks. He borrowed the term from a line in James Joyce's 'Finnegan Wake'—"Three quarks for Muster Mark ..."

Some scientists believe that the discovery of quarks may represent the end of scientists' search for the most basic unit of matter. Quarks exist only in theory. Physicists try to prove the existence of quarks by producing them in [thomic] smashers. In theory, quarks carry electrical charges. An electrical charge is a property of the electrons and protons. Having electrical charges to emit from themselves is an indication that quarks, even if they exist, cannot be the basic particles of matter as these particles must have these qualities of absolute solidity and indivisibility to give permanence to matter. Having these qualities make the basic particles of matter incapable of emitting or absorbing anything.

I will re-emphasize this very important fact: The only possible particles that can be the basic particles of matter are the absolutely solid, indivisible atoms. The irrefutable evidence for their existence comes from the impossibility of infinite divisibility of matter. Their existence is not amenable to scientific check because they exist far below the range of the senses and individually without color. They cannot be seen with the most powerful microscope. *Knowledge of their existence comes from the intellect, not from the senses.*

By now it should be clearly understood that modern physicists, as I maintain, mistake the thom for the atom and that this serious blunder has prevented them from understanding that:

1. Atoms are the smallest particles of matter that can exist—the ultimate and smallest division of matter.

2. Nothing exists but atoms and kenon.

3. Everything is made up of atoms and kenon, including all subthomic particles—electrons, protons, neutrons, etc.

4. Everything made up of atoms is an aggregate of atoms.

5. Any aggregate of atoms can be split or divided, but not atoms themselves.

6. Atoms have *dynamis* from their sheer existence.

7. The *dynamis* of atoms is their ability or capacity to do, act, or produce a certain effect.

8. *Energeia* is the Greek name for the activity of atoms or aggregates of atoms doing work, such as producing heat, light, etc.

9. Some of the tiniest aggregates of atoms, tightly compacted, have tremendous explosive power.

10. When aggregates of atoms collide with other aggregates of atoms, moving at a high rate of speed, they tend to explode and release heat.

11. The heat possessed by an object is due to the vibration of its atoms.

12. Atoms enclosed in compounds vibrate constantly and rapidly, producing heat.

13. Being made up of vibrating atoms, all matter can give off heat.

14. The properties and behavior of matter arise from the properties and behavior of its atoms.

15. Everything about matter can be explained in terms of the movements of its constituent atoms.

16. The interaction of atoms ultimately causes all the different phenomena in nature.

17. The proper way to do physics is to study the interaction of atoms.

18. Atoms are the only forces (dynameis) in the universe.

19. Gravity and repellency are fundamental properties of atoms.

20. Atoms attract each other when they are little distance apart, but repel each other when they come too close to each other.

21. All of nature can be understood by describing the size, shape, solidity, *varos* (weight or force of mass), gravity, motion, position, arrangement, repellency, and dynamis of atoms.

22. The original classical mechanistic goals of explaining all of nature in terms of the motions of atoms must be re-embraced.

I will now end this discussion with the following factual explanation. Atomism, or the Atomic Theory, gives the final answers to the questions, "What is the universe made of?" and "How does it work?" It is the starting point from which to develop an understanding of everything knowable. It is the only theory that can explain change and multiplicity. It cannot be replaced by the theory of thoms, or Thomic theory, especially since thoms are not the basic particles of matter and cannot be used to explain the behavior of subthomic particles, such as the electrons, protons, neutrons, quarks, and atoms. The "so-called" atomic theory of modern physicists claims that all matter is made up of atoms, even though subatomic particles of matter cannot be. Scientists must embrace the original and "true" Atomic theory if they want to truly understand the properties and behavior of matter. However, they may continue to make use of the Thomic theory to explain the properties and behavior of the smallest unit of an element that retains the properties of that element, not forgetting that thoms, as well as all their sub-thomic particles are ultimately aggregates of atoms.

According to the Thomic theory:

1.   All elements are made up of thoms.

2.   An element is a substance made up of just one kind of thom.

3.   A thom is the smallest unit of an element that retains the properties of that element.

4.   All thoms are composed of atoms.

5.   Atoms can neither be created nor destroyed.

6.   The thoms in a substance make up its mass.

7.   Thoms of the same element always weigh the same, and they are also alike in their other characteristics.

8.   Thoms of one element are different in weight from those of other elements.

9.   Thoms of different elements combine to form compounds.

10. Thoms, being compounds of atoms, are "splittable."

11. Subthomic particles, such as electrons, protons, neutrons, quarks, and atoms, do not retain the properties of an element.

Always remember that a satisfactory theory explains the known facts.

I will now end this essay with this final explanation of the facts known about atoms and thoms. Atoms and thoms are what basically exist. All solid objects and liquids and gases are nothing more than complex arrangements of atoms and thoms.

Essay #3
A Refutation of the Big Bang Theory

Let us examine the Big Bang theory of the origin of the universe, a theory that I will refute as being pseudo-science. This theory claims that the universe originated as an infinitesimal hot dense mass that exploded about 10 billion to 20 billion years ago, throwing out matter in all directions, from which galaxies and stars formed. Big Bang theorists say that all the matter that existed was concentrated in that infinitesimal mass. They do not say where the mass came from and what made it explode. They maintain that before the Big Bang event there was no matter, no space, no time, and no energy. They say that all of these things came into existence at the moment of the Big Bang event. They say that matter and energy came into existence out of the void, a place containing nothing.

The idea that matter and energy came into existence out of nothing is "naïve" and "unscientific." In reality, nothing ever comes into existence out of nothing. Observation of nature shows that when things come into existence, they come from things already existing. Since matter is the ultimate substance that makes up the material universe, and since nature shows that matter can not be created or destroyed, I believe in the eternity of matter.

Since matter has always existed, it has always occupied space and has never existed outside of eternal time. As for energy, it is produced solely by the movements of matter. The word *energy* is taken from the Greek word *energeia*, the name for "activity" of matter. Energy exists "prior" to any non-eternal event. Hence, no Big Bang event created it either.

Obviously, the Big Bang theory is a false theory of the origin of the universe. But, let us examine the evidence that Big Bang theorists offer to support the theory. They say that the strongest evidence for the theory comes from observation of faint cosmic background radiation coming from all directions in space. They believe that this radiation is all that remains of the Big Bang explosion. They say that the temperature of the radiation is almost exactly the same in all directions.

Radiation from such an enormous explosion as the Big Bang would be coming from widely separated parts of the universe and certainly would not be so uniform. This fact goes against the Big Bang theory, instead of supporting it.

Many Big Bang theorists believe that the universe is expanding and that it will eventually stop expanding and then collapse. They believe this because it is being observed through optical telescopes that galaxies are systematically speeding away from each other at ever-increasing speeds. The theorists think that the galaxies' speed of movement will eventually abate causing the universe to collapse. They call this theory the "Big Crunch."

I said that I would refute the Big Bang theory as being pseudo-science. Real science is based on the reality that the universe is made up of atoms and empty space, and that it works atomically. Let it be understood that atoms are the smallest particles of matter that can exist—the ultimate and smallest division of matter. Let it also be understood that, since it is impossible for atoms to come into existence out of nothing or pass away into nothing, they are eternal and indestructible. There are infinitely many atoms in existence, scattered throughout the infinite universe. The interaction of these tiny atoms ultimately causes all the different phenomena in nature. Atoms have come together by chance to form our world and all the things in it, as well as innumerable other worlds. Our world, like the innumerable other worlds, is an accidentally generated aggregate of atoms of finite duration.

Big Bang theorists think that atoms are particles of matter that were formed after the Big Bang event. They say that at the time of the Big Bang event, the temperature of the hot dense mass that exploded was about 100,000,000,000 degrees Celsius (10" °C), and that several hundred thousand years later had dropped to a mere 2,500 °C. This temperature was cool enough to allow electrons and nuclei to join up and form the first complete atom. They describe the atom as being a tiny particle that resembles a tiny solar system with electrons whirling around a nucleus of protons and neutrons in fixed circular orbits. Electrons whirl around the nucleus of protons and neutrons at fantastic speed, completing billions of trips in a millionth of a second. An average atom is about eight billionths of an inch across (20 billionth of a centimeter). A billion atoms in a row would make a line a few centimeters long. It would take about 2,500,000 atoms placed side by side to stretch across the head of a pin. A drop of water contains more than 100 quintillion atoms. There are more atoms in one grain of sand than there are leaves on all the trees in the world.

Protons and neutrons are about 100,000 times smaller than atoms. They make up about 99 percent of an atom's mass. A proton has about 1.673×10^{-24}

grams; a neutron has about 1.675×10^{-24}; and an electron has 9.110×10^{-28}. Protons, neutrons, and electrons are made up of smaller parts called "quarks," which are at least 1,000 times smaller than protons.

The particle described by Big Bang theorists as being an atom is certainly not an atom. As I have already explained, atoms are the smallest particles of matter that can exist—the ultimate and smallest division of matter. It was explained 2,400 years ago by the Greek Atomist Leucippus, who originated the idea of the atom, if matter is repeatedly cut up, the end result will be uncuttable pieces of matter, "atoms." The word *atoms* comes from the Greek word *ātomos*, meaning "uncuttable." If the description of a particle does not fit this definition of an atom, it cannot be an atom.

I maintain that the universe has always existed, and that it is made up of an infinite number of eternal and indestructible atoms scattered throughout its infinite space. The atoms are the ultimate elements from which nature forms, increases, and sustains all things, and into which she again resolves them when they perish. The Atomists often call them "matter" or "raw material" or "generative particles of things" or "seeds of things;" and since they are the ultimate constituents of all things, they are also called "ultimate particles" or "primary particles."

Atoms have come together by chance to form our world and all the things in it as well as innumerable other worlds (planets and galaxies). The universe was not created 10 billion to 20 billion years ago from the explosion of a hot dense mass that contained all of the matter which existed, and was smaller than the nucleus of a thom. If all of the atoms that make up the universe had ever been concentrated in a single mass, that mass would have been infinitely huge because atoms are absolutely solid particles. Being squeezed together would not have changed the size of a single atom.

The infinitely vast universe contains billions of huge galaxies that are separated from one another by millions and billions of light years. Traveling in a space ship at the speed of light (186,00 miles [300,000 kilometers] per second), it would take 100,000 years to cross our own Milky Way galaxy from one side to the other. It would take 2 million years traveling at the same speed to reach the nearest other galaxy, *Andromeda*. Beyond this neighboring galaxy are billions of "islands" of galaxies separated by billions of oceans of space. There are *quasars* (bright nuclei of galaxies) appearing to be as far away from the Earth as 14 billion light-years, and receding at speeds approaching that of light. All of the galaxies contained in the infinite universe are swarm centers of whirling atoms or aggregates of atoms that are in perpetual motion and are receding from one another under the effect of the whirling motion and gravitational pull on them by older and more massive galaxies existing great distances away from them.

Motion is a fundamental property of the atoms. It is eternal and indestructible just as are the atoms themselves. The ever-shifting arrangement of the self-moving atoms in space is the ultimate reality behind all appearances. Their interaction in space ultimately causes all the different phenomena in nature.

As for the Big Bang theorists' idea that the universe will eventually collapse because the galaxies will stop expanding because the force of the Big Bang explosion that created them will eventually abate, I maintain that this incredibly vast universe of which we have only a sectional view was not created by a Big Bang explosion from a hot dense mass amounting to nothing. The atoms that have come together to form the galaxies will eventually dissipate, causing existing galaxies to disappear into the void, but this does not mean that the eternally existing infinite universe will collapse. The atoms of which it is made cannot be destroyed and will go on forming other galaxies throughout the universe.

By now, it should be crystal clear to you that the Big Bang theory of the creation of the universe is false and pseudo-science.

Essay #4
A Refutation of the String Theory of Matter

"String theory" is a systematic statement of principles involved of equations in mathematics, proposing that matter is made of strings. In physics, according to *Random House Webster's Unabridged Dictionary*, a "string" is a mathematical entity used to represent elementary particles, as gravitons, quarks, or leptons, in terms of small but finite stringlike object existing in the four dimensions of space-time and in additional, hypothetical, spacelike dimensions. The theory of such objects (string theory) avoids the many mathematical difficulties that arise from treating particles as points.[23]

The *Oxford Dictionary of Physics* defines a string as [a] one dimensional object used in theories of elementary particles and in cosmology (cosmic string). String theory replaces the idea of a pointlike elementary particle (used in quantum field theory).[24]

According to string theory, strings are the smallest possible particles, with a length of 10^{-33} centimeters, no width and no height. They can be open with two ends, or closed into a circle. A string has a fixed point in time and space.

Let it be understood that strings are mathematical entities, not material entities; hence, nothing in reality is made of strings. A string cannot represent the smallest possible particle because the smallest particle is the "atom," which is a three-dimensional object, not a one-dimensional object. In reality, there is no such thing as a one-dimensional object. All objects are three-dimensional, having length, width, and height. All elementary particles as gravitons, quarks, or leptons are made of atoms, which are not stringlike objects.

String theory asserts that there is an atomic structure to space, but ignores the fact that atoms, being three-dimensional objects cannot form themselves into one-dimensional particles. I ask string theorists, if one-dimensional strings are the fundamental objects in the universe, how do they join themselves together to form three-dimensional objects? They can only join with themselves to extend their length. They cannot give themselves width and height and, therefore, cannot form themselves into three-dimensional objects.

I will reiterate: Nothing in reality is made of strings. Strings are not real. Everything is made of three-dimensional atoms.

Obviously as a theory of matter, the string theory is fatally flawed.

Essay #5
Why Physicists Do Not Know What Energy Is

Energy is a basic concept in modern physics. Physics is the science that studies what things are made of and how they work. In modern physics it is taught that the universe and all things in it are made of matter and energy, and that energy is what changes and moves matter. Matter is defined as anything that has mass and occupies space. Mass is defined as the amount of matter that makes up an object. Energy is defined as the ability to do work, such as moving things or giving heat or light. The word "energy" was introduced into the technical vocabulary of science in 1807 by the English physician and physicist Thomas Young, who was a professor of Natural Philosophy at the Royal Institution of London. He defined energy as the ability to do work. Be cognizant that this definition does not tell us what energy is, only what it does or can do. As the American theoretical physicist and Nobel Prize winner Richard P. Feynman (1918-1988) explained, "It is important to realize that in physics today, we have no knowledge of what energy is."[25]

Why don't physicists know what energy is? They don't know what it is because they think that it is something that it is not. The word *energy* comes from the Greek word *energeia* (from *en*, in, plus *ergon*, work, and from *energos*, active), a word used by ancient Greek philosophers to refer to the "activity" of a thing directed toward work, or the accomplishment of something. When Young introduced the word energy into the technical vocabulary of science, he defined it incorrectly. He gave it the definition of the Greek word *dynamis*[26], which means the ability or capacity to do something or to produce a certain effect. An object has dynamis from its sheer existence, but not energy. It is the dynamis of an object that makes energy possible.

Another reason why physicists do not know what energy is is because it exists only as the movements of atoms and aggregates of atoms. Present-day physicists are not acquainted with these atoms. They think that atoms are the tiny things that they call atoms in modern physics, which are splittable into subatomic particles. Those tiny things are not atoms, they are "thoms." The word "atom" means "uncutttable." If the description of a thing does not fit this definition of an atom, it is not an atom. Present-day physicists claim that energy is not composed of atoms, thinking of the thoms that they mistakenly consider to be atoms. They do not understand that nothing can come into existence without atoms, and that energy is produced by the movements of atoms.

Physicists claim that energy cannot be created or destroyed, but since energy is created by the movements of atoms, that is not true. Let it be understood that all aggregates of atoms are objects of atoms carrying on energy. The energy of objects varies in duration. The energy carried on by an object is created by the movements of the atoms of which it is composed. When the atoms of an object dissipate, and thereby destroy the object, the object's energy is destroyed.

In contradiction to their claim that energy cannot be created or destroyed, physicists claim that energy can sometimes be created by destroying matter. Matter cannot be destroyed, but only broken up into constituent atoms. Its atoms can re-arrange themselves and carry on different kinds of energy.

Physicists claim that atoms can be split to release an enormous amount of energy. This claim is not true either. Atoms are the smallest particles of matter that can exist—the ultimate and smallest division of matter. They are absolutely solid and cannot be split. No object can release energy because no object is composed of energy. An object can only release some of the atoms of which it is composed. The tiniest mass of atoms, traveling at the speed of light—186,000 miles (300,000 kilometers) per second—carries a tremendous amount of dynamis (power). Even a single atom, traveling through the kenon (empty space) of the universe faster than the speed of light, carries such dynamis.

Physicists claim that heat, light, sound, wind, and electricity are forms of energy. Again, they are wrong! These are *effects* of the energy of atoms. Some physicists will undoubtedly argue that energy produces these effects. I will maintain that it is not the energy of atoms that produces effects; it is the atoms that act and produce effects. If one doubts this, one need merely to understand that the energy of atoms cannot be detached or separated from the atoms.

Essay #6
What Makes Matter Change and Move?

Matter, the "stuff" from which things are made, is constantly changing and moving. All changes in matter are the result of atoms moving and combining in different ways. All matter is made up of atoms. Atoms are the smallest particles of matter that can exist—the ultimate and smallest division of matter. All atoms, even those in compounds, are in perpetual motion.[27]

There are four causes of an atom's motion. The first is the *dynamis* (force) of its own *varos* (weight, or mass) pressing down upon the *kenon* (pure empty space) that separates atoms and in which atoms move. This kenon cannot support varos and hence must allow an atom to fall straight downward, as its own nature demands. This starts the motion of an atom, which Atomists call the "original motion" of an atom. Atoms fall through the kenon at equal speed faster than light—186,000 miles (300,000 kilometers) per second, since in kenon there is nothing to slow them down.

Some people claim that the idea that atoms fall downward through the kenon of the universe is naïve, thinking there can be no "up" or "down" in the infinite kenon of the universe. They fail to grasp the fact that the movement of the atoms from one place to another in the kenon is *directional*, and that the varos of the atoms only allows them, initially, to move themselves in one direction—through the kenon existing below them. Movement in this direction is "downward," movement in the opposite direction would be "upward."

The second cause of an atom's motion is the *parenklisis* (swerve), whereby an atom may shift from its straight downward path onto an adjacent one under the influence of gravity. Gravity is the force that atoms exert on each other because of their mass. Every atom has its own gravity and attracts every other atom in the universe. Gravity's strength depends on an atom's mass or amount of matter. More mass means stronger gravity. Thus, the size of the atom is important—a large atom attracts other atoms more strongly than a small one. A large atom pulls small atoms toward it. The greater the distance between the atoms, the weaker the gravitational pull between them. When atoms are traveling close enough to each other, a large atom pulls small atoms off course, onto an adjacent one. This happens at no fixed place and time, meaning that the occurrence is random. It causes atoms to collide with other atoms falling on adjacent courses.

The third cause of an atom's motion is collisions with other atoms. Sometimes atoms just bounce apart in opposite directions. At other times, under the influence of gravity, they are drawn into a whirl or *vortex*, in which atoms of a similar shape and size come together. In this vortex the small atoms are squeezed out into the outer kenon, but the remainder tend toward the center, where they form a spherical mass. More atoms are drawn into the mass on contact with the whirl, and some of these are found to cohere by the speed of the revolution, thus forming the celestial bodies.

The fourth cause of an atom's motion is its dynamis (power, or force) or repellency, which is an opposing force to gravity. Just as atoms attract each other when they are a short distance apart, they drive or force back, or hold or ward off, one another when squeezed into one another. Their dynamis of repellency keeps atoms separated from one another and contributes to their perpetual motion.

Notice that I have explained what makes matter change and move without mentioning energy. Physicists who are energy theorists claim that energy is what makes matter change and move. They are wrong! They define "energy" as "the ability to do work," such as changing the shape or the phase of matter, or moving matter from place to place, or giving heat or light, or making living things grow. Their definition of energy is incorrect. The word *energy* is taken from the Greek world *energeia* (from *en*, in, plus *ergon*, work, and from *energos*, active). Energeia is the Greek word for activity. Physicists confuse it with the Greek word *dynamis* (power, force, or strength), which means the ability to do something or produce an effect. The energy, or activity, of matter is produced by the dynamis of its moving atoms. Dynamis and motion are fundamental properties of the atoms. An atom possesses dynamis from its sheer existence. All changes in the universe are reducible to the dynamis of the movements of the atoms in the kenon. Without this dynamis, atoms would be unable to move and there would be no energy, or activity, of things in the universe.

Let it be understood that it is not energy that makes matter change and move. The movement of matter always exists prior to the energy it produces. Energy is not an entity that exists separately from matter. It is always equivalent to the rate at which matter does work.

Essay #7

Einstein Was Wrong: $E \neq mc^2$

Let me begin this essay with a remark attributed to the great German philosopher Arthur Schopenhauer (1788-1860): "All truth passes through three stages. First it is ridiculed. Second, it is violently opposed. Third, it is accepted as being self-evident."[28] I believe that this remark will apply to this essay.

In 1905, the German-born theoretical physicist Albert Einstein (1879-1955), who became an American citizen in 1940, theorized that matter and energy are equivalent and that matter can be changed into energy and energy into matter. What is matter? What is energy? Matter is the ultimate substance that makes up the physical or material universe. In physics, matter is defined as anything that occupies space and has mass (or weight), which is the influence of gravity on mass. Mass is the amount of matter in an object. All matter is made up of atoms. Atoms are the smallest particles of matter that can exist—the ultimate and smallest division of matter. If matter is repeatedly cut up, the end result will be uncuttable pieces of matter, atoms. The word *atom* comes from the Greek word *atomos*, meaning uncuttable." Matter cannot be destroyed into nothing, but only broken up into constituent atoms. Matter ordinarily exists in three forms: solids, liquids, and gases. Although it cannot be created or destroyed, it can be changed from one form to another. For example, water can be frozen into ice or boiled into vapor.

In physics, matter is distinguished from energy, which it claims causes matter to change and move, but has no mass of its own. Energy is defined as the ability to do work, such as moving things or giving heat or light. Another way energy is defined is the ability to change the shape or the phase of matter. It is also defined as the ability of living things to grow. Notice that none of these definitions tell us what energy is but only what it supposedly does or can do. As the American theoretical physicist and Nobel Prize winner Richard P. Feynman (1918-1988) explained, "It is important to realize that in physics today, we have no knowledge of what energy is."[29]

All of the definitions of energy given in modern physics are wrong. The word *energy* comes from the Greek world *energeia* (from *en*, in, plus *ergon*, *energos*, active), a word used in philosophy to refer to the "activity" of anything directed toward work, or the accomplishment of something. Energy is always equivalent to the rate at which a thing does work. In 1807, an English physician and physicist Thomas Young (1773-1829), who had been named professor of natural philosophy at the Royal Institution of London in 1801, wrote a paper in which he wrongly defined energy as the ability to do work. This definition is the definition of another Greek word, *dynamis*, meaning power, force, strength. The word *dynamis* comes from the Greek word *dynasthai*, meaning "to be able." Plato used it to mean: (1) the power of transferring or imparting activity or change to something else, (2) the ability to receive power (force, motion, activity), and (3) the capacity to be influenced by forces (change movement, etc.)[30]. Aristotle used it to mean (1) that power which causes change and/or (2) that state of potential a thing has to produce change or to become other than what it is.[31]

Atomists use the word *dynamis* to mean the ability of atoms or aggregates of atoms to do something or produce a certain effect.

It must be understood that dynamis (power) is the ability of a thing to do work, while energy is the activity of a thing directed toward work. Further, an object possesses dynamis by its sheer existence, but not energy. The dynamis (power) of an object is "carried" by its motion. It is called the impact. An object moving at high speed has high impact if it strikes another object. The greater the mass and speed of an object, the greater its impact. The greater the dynamis carried by objects, and the greater the speed at which they move, the greater is their energy (rate of work).

Dynamis can be calculated by using several different formulas:

Calculating dynamis from force:	Dynamis = Force x Speed
Calculating dynamis from energy:	Dynamis = Energy/Time
Calculating dynamis from work:	Dynamis = Work/Time

I will now discuss Einstein's famous equation E = mc², which he used to express the idea that matter can be changed into energy. The symbols in the equation stand for energy (E), mass (m), and the speed of light (2). The "2" means that the speed of light is multiplied by itself (or squared). In Einstein's equation, c is equal to 186,000 miles (300,000 kilometers) per second, and therefore, c² is equal to 34,596,000,000 miles² per second². This is a large number.

When the value of c² is multiplied by the amount of matter, "m", the result is a fantastically large value for the energy released. What Einstein's equation means, then, is that a miniscule amount of matter can be converted into an enormous amount of energy, which is not true.

There are several problems with Einstein's equation $E = mc^2$. First of all, it is based on the wrong idea that matter is equivalent to something that is not matter, which is impossible because nothing exists but matter and kenon. Second, it is based on a mistaken idea about what energy is. Third, it is based on the wrong idea that energy cannot be created or destroyed, when, in fact, all energy is created and varies in duration. Fourth, it is based on the wrong idea that energy represents a certain amount of matter, resulting from matter, or mass, traveling at the speed of light, squared. *Energy, being the activity of matter directed toward work, is a manifestation of the motion of matter moving at any speed while doing work.*

Here, let me explain that energy does not equal mass traveling at the speed of light, squared, but rather *work divided by time (E=work/time).*

Einstein claimed that, according to the equation $E = mc^2$, atoms could be split to release enormous amounts of energy, which is impossible because atoms cannot be split. As I explained earlier, atoms are the smallest particles of matter that can exist—the ultimate and smallest division of matter.

I acknowledge the fact that Einstein's equation $E = mc^2$ led to the splitting of the "so-called" atom and the development of the use of the "so-called" atomic bombs that destroyed the Japanese cities of Hiroshima and Nagasaki in 1945. The splitting of the "so-called" atom produces "so-called" atomic power, but that does not make the equation $E = mc^2$ correct. The fact remains that the equation says what is **not** true. Energy is not equal to matter, or mass, traveling at the speed of light, squared.

I will reiterate: energy, being the activity of anything directed toward work, is equal to work divided by time (E=work/time). Work equals force times distance (W=fxd). Force equals mass times acceleration (F=ma).

I am sure that had Einstein understood that all matter is made up of eternal and indestructible atoms, he would not have claimed that an atom could be split, only an "aggregate of atoms." Had he understood that atoms have dynamis, but not energeia, he would not have claimed that even an aggregate of atoms could be split to release energy. He would have understood that many atoms make up all matter, and that moving atoms have *momentum* (quantity of motion), which he confused with energy.

Momentum is a measure of an atom's or group of atoms' mass in motion; the momentum of an atom or group of atoms is the product of its mass and its velocity. To calculate an atom's or group of atoms' momentum, you can use the following formula:

$$\text{Momentum} = \text{mass x velocity}$$
$$p = mv$$

In this equation, p stands for momentum, m for mass, and v for velocity. In standard units, the mass of an object is given in kilograms (kg), and the velocity is given in meters per second (m/s). Therefore, the unit of momentum is the kilogram-meter per second kg m/s). The unit of momentum combines mass, length, and time.

Let it be understood that high velocity of an atom or tiny group of atoms with very tiny mass can produce a large momentum. This large momentum is what Einstein wrongly took to be energy, instead of dynamis. When the tiniest aggregate of tightly packed atoms is split, its atoms will fly apart with tremendous explosive dynamis (power).

I hope by now I have convinced you that Einstein's theory that matter and energy are equivalent and that matter can be changed into energy and vice versa is wrong. The theory has led to a false restructuring of the laws of physics and to an erroneous conception of the universe. The theory should be abandoned, and the theory that energy is equal to work divided by time (E=w/t) should be embraced.

Essay #8
A Refutation of Einstein's Description of Light

Albert Einstein described light as being composed of *photons*—tiny particles with no mass. Being an Atomist, a person who believes that everything is made of atoms and empty space, and that nothing exists but atoms and empty space, I challenge Einstein's claim. *There is no such thing as a particle without mass.* Physicists describe many particles as having "zero mass," which is impossible. A particle is a tiny bit of matter. Matter is anything that occupies space and has mass. Mass is the amount of matter in an object. I will reiterate: *There is no such thing as a particle without mass.* Every particle has mass, no matter how infinitesimal it may be. Particles with no mass, photons, do not exist; therefore, light is not composed of them.

Light is a radiant or bright *effect* caused or produced by the interaction of an aggregate, or quanta, of massy atoms. An aggregate is a number of distinct things gathered or clustered into a whole or mass. The word *quanta* is the plural of *quantum*, which comes form the Latin word *quantus*, meaning "how much." A quantum is a very small, indivisible quantity or unit of something. For light or any other electromagnetic radiation, one quantum is equal to its frequency multiplied by a particular number that never changes. An atom is the smallest particle of matter that can exist—the ultimate and smallest division of matter. If matter is repeatedly cut up, the end result will be uncuttable pieces of matter, "atoms." The word *atom* comes from the Greek word *atomos*, meaning "uncuttable" or "indivisible." Here, let it be understood that the "splittable" atom of modern physics is a *misnamed* particle, and that I am not talking about the atom, which is made up of "so-called" subatomic particles-protons, neutrons, and electrons. The interaction of the atoms that I am discussing in this essay ultimately causes all the different phenomena in nature, including light.

Light comes from the sun and travels at 186,000 miles (300,000 kilometers) a second. The sun is made up of an infinite number of atoms. It provides Earth with most of its heat and light. The sun has a diameter of about 865,000 miles (1,392,000 kilometers) and is about 93 million miles (150 million kilometers) from Earth. Its temperature is about 10,000 degrees Fahrenheit (15,000,000 degrees Celsius) at the center. Light from the sun takes about 8 minutes to reach the Earth.

Einstein called light a form of energy. In physics energy is defined as *the ability to do work.* That definition is a deviation from the etymological meaning of the word, and it creates ambiguity. The word *energy* comes form the Greek word *energeia* (from *en*, in, plus *ergon*, work, and from *energos,* active) meaning "the activity of anything directed toward work." I only accept the etymological mean- ing of energy, judging it to be the only appropriate meaning of the word. The energy of things varies in duration. Energy is not the ability to do work. Energy, being the activity of atoms, cannot exist apart from atoms.

Now physicists must acknowledge that light is composed of atoms and, there- fore, does not exist as something *immaterial.* Nothing can exist which is immate- rial, except the vacuum of empty space. Since light takes up space, we know that it has mass. Mass is the number of atoms that make up a material object. Material objects are of two kinds: atoms and aggregates of atoms. Light is a quanta of aggregates of atoms.

Being composed of atoms, light occupies space and has mass. Mass is the number of atoms that makes up an object. Since light has mass, it also has weight, but its weight is infinitesimal; that is, it is so small or insignificant as to be close to nothing.

Here, let it be understood that even though light is infinitesimal in mass and weight, it is still a form of matter. Nothing can exist which is immaterial, except void (empty space).

Since light is made up of material atoms, when objects absorb light they increase their atomic density or size.

Light penetrates objects through their pores. The word *pore* comes from the Greek word *poros*, meaning "opening." Pores are tiny openings, as in the skin or a leaf, serving as an outlet or inlet. Perspiration escapes through the pores in the skin. Pores are also spaces between grains in rock or soil. Everything, except atoms, has a porous texture.

Light cannot penetrate atoms. This is because atoms are absolutely solid. Being absolutely solid, light bounces off of them in all directions.

Physicists who study the behavior of light have learned that light travels in waves, just as heat and sound do. Heat and sound, like light, are composed of aggregates, or quanta, of rapidly moving atoms. Physicists have measured light waves and have found that they are very short, only a very small fraction of an inch (millimeter) long.

One may wonder about the size of an atom and how many it takes to make an inch (or millimeter). Since an atom is an elementary particle; that is, a particle that cannot be divided into parts, and since physicists claim that an electron is such a particle and would be the largest known atom, I will say atoms range in size from that of the electron to less than any specified number, as close to zero as possible. An electron has a diameter of 10^{-12} m. This diameter of an electron is written with a decimal point followed by eleven zeros and a 1 (0.000000000001). This means that it takes approximately one billion (10^9) electron size atoms strung together to equal one millimeter (0.03937 inch), which is about the thickness of a stack of ten sheets of paper.

Personally, I reject the idea that an electron is an elementary particle. This is because electrons traveling in a cathode ray near the speed of light increase in size. A true elementary particle is an atom. An atom is absolutely solid and is, therefore, immutable. Being immutable, an atom's size cannot change. The fact that an electron can increase in size proves that it is not an atom and is, therefore, not an elementary particle.

When the sun gives off light waves, tiny aggregates of atoms radiate from it in all directions. We see these tiny aggregates of atoms as rays of light. By knowing that light is composed of tiny aggregates, or quanta, of massy atoms, and by learning about the properties of atoms, physicists will be able to develop a better understanding of the properties and behavior of the aggregates of atoms that produce light, and how light affects other material things.

Essay #9
A Refutation of Einstein's Theory of Time

In 1905, Albert Einstein stated as part of his *special theory of relativity* that *time is relative*. He described time as speed measured by clocks and maintained that time is not constant throughout the universe. He claimed that as objects approach the speed of light—186,000 thousand miles (300,000 kilometers) a second—time slows down. Time, he said, passes more slowly for them. He was wrong and let me explain why.

First, let me explain what time is, because Einstein's description of time is wrong. *Time is the eternal and infinite continuum of the existence of the eternal and infinite universe.* By universe, I mean the totality of all matter and space that exists. Matter is the ultimate substance that makes up the material universe. Since it is impossible for matter to come into existence out of nothing or pass away into nothing, it is eternal and indestructible. Space is the whole expanse of the universe. Like matter, space is eternal and infinite.

It is very important to understand that matter is bounded by empty space and that empty space is bounded by matter. This makes the universe an infinite continuum of alternating matter and empty space.

Now, let it be understood that time has no independent existence from the eternal and infinite universe: rather it is an inherent property of the universe. If the universe did not exist, time would have no existence. Since time has no independent existence from the eternal and infinite universe, the definition that I have given of time is certainly appropriate: *Time is the eternal and infinite continuum of the eternal and infinite universe.*

Being the eternal and infinite continuum of the existence of the whole universe, time is the *same* throughout the universe. It has flowed without change from eternity. It passes at the same rate for everything existing in the universe. It cannot be interrupted; therefore it is *constant* throughout the universe. It flows in a constant, absolute straight-forward direction from eternity with the whole universe. Now it is easy to see that Einstein's claim that time is relative is wrong.

Now, let me address his claim that as an object approaches the speed of light, time slows down. This claim is also wrong. Suppose two objects come into existence at the same time and one remains almost motionless while the other streaks continuously across space at the near speed of light, and they both go out of existence at the same time. The passage of time will be at the same rate for both. This is proved by the fact that they both had the same duration of existence. The speed of travel of an object has no affect on time. Time remains constant. It never slows down.

Let me return to Einstein's description of time as speed measured by clocks. Clocks measure the speed of change occurring during the passing of time, not time itself. Unlike time, change has a beginning and an end and, thus, can be measured. It can speed up, slow down, and stop. Time itself exists before change, while change is happening, and after change ends.

A good way to understand that the clock is a standard of change—a means by which we measure other changes—is to relate the change (or movement) of a runner to the change of the clock. It is a correlation, or ratio. We summarize the number or rate of change by saying, "The runner ran the mile in four minutes and thirty seconds." We do not relate the change of the running to time but to the clock.

Einstein *mistook* change for time and wrongly chose the clock as a standard of time. He used the clock to support his theory that time is relative, claiming that time is represented by the change or movement of the clock. He said that clocks approaching the speed of light out in space would slow down, therefore time would slow down. Based on this assumption, time would stop if the clock stops and we would live in a timeless universe.

Obviously, Einstein's theory that time is relative is untrue. It contradicts the reality that eternal and infinite time is the same throughout the universe. Further, it leads us astray from arriving at a correct cosmological conception of the existence of the universe and how the universe works.

Essay #10

A Refutation of Einstein's Theory of Gravitation and Curvature of Space

In 1915, Albert Einstein said in his general theory of relativity that gravitation is a property of the space in which bodies exist. In physics, gravitation is a natural force that attracts all objects in the universe to one another. Every piece of matter in the universe attracts every other piece of matter. The amount of matter is important: a large piece attracts other matter more strongly than a small piece. The amount of the gravitational force pulling the bodies (masses) together depends on two things: the mass of bodies and the distance between them. The greater the distance between the bodies, the smaller the gravitational pull between them.

Let it be understood that gravitation is a property of all the bodies in the universe, and that a property is something that cannot exist apart from a body. Einstein looked upon gravitation not so much as a property of all bodies but wrongly as a property of the space in which bodies exist. Space is a void and a void possesses no qualities whatever, no powers, no potentiality, no tangibility in any way. It is impossible for gravitation to exist in a void. Therefore, wherever there is gravitation, there are bodies. The bodies may be so small that they cannot be seen. For example, the atoms that make up all the matter in the universe are such bodies.

Einstein held that what appears to be a gravitational force between bodies is a kind of curvature of space. He said that the gravitational force between bodies creates a sort of cosmic suction that curves space around the bodies. This is an incredible idea, especially since space is a void (nothingness). It is impossible for space to curve. Only objects can curve, in form and movement.

Einstein thought that gravity should be explained as a curvature in the geometry of space, not as a force. I will reiterate: *He thought of gravity not so much as a property of all bodies that causes them to attract all other bodies but wrongly as a property of the space in which bodies exist.* Einstein realized that gravity is a property of matter, but he thought that the effect of the gravity of material objects is to "curve" the space around them. He did not understand that there is a field of invisible objects surrounding the visible objects, and that it is this field of invisible objects that "curves" around visible objects. The space in which objects exist does not curve.

Einstein had another wrong idea about gravitation. He held that its effects much travel through the universe at a slower speed than light (186,000 miles [300,000 kilometers] per second). He held the speed of light to be the upper limit of all speeds. He did not understand that the atoms—the tiny particles of matter that make up the universe—travel many times faster than the speed of light. Since all atoms possess the property of gravity, and since gravity cannot be detached from them, their gravitational effects must travel as fast as they do.

Einstein claimed that gravity could slow down time. He did not understand that time is the entire period of existence of the duration of the eternal universe. Time passes with the constancy of the existence of the universe. The idea that gravity can slow down time is preposterous.

I believe that I have given sufficient information to show that Einstein's theory of gravitation and curvature of space is definitely wrong.

Essay #11
Einstein's False Conception of the Universe

Albert Einstein's conception of the universe was deeply flawed with shallow reasoning and faulty logic. To understand that this is so, we must first understand what is meant by the word *universe*. Universe means the totality of all the matter and space that exist; no matter or space can exist outside the universe. The universe is infinite in all directions. The Atomists simply call it "the all."

Einstein conceived of the universe as being finite in space but infinite in time, remaining the same fixed size for eternity. All of the existing matter was taken to be distributed in a uniform and continuous way throughout all of the finite space. He held that the universe has no spatial boundaries, and that it curves back on itself like a circle or folds back on itself like a blanket. He also held that the universe is static. He did not believe that the universe could be in motion.

Einstein's conception of the universe as being finite in space is at odds with reality. We ourselves observe that nature compels matter to be bounded by void (empty space), and void by matter, so that by their alternation it makes the universe infinite in matter and space. As far as Einstein's conception of the universe as being infinite in time, I agree. I understand that the universe has always existed and hence has never existed independently of time. Time itself is the eternal and infinite continuum of the existence of the eternal and infinite universe.

I agree with Einstein that the universe remains the same fixed size. I understand that the universe is infinite, and that the infinite does not change its size. The infinite extends beyond size and measure.

Since I understand that the universe is infinite in space, I reject Einstein's idea that the universe curves back on itself like a circle. Being infinite in all directions, the universe contains all of the space that exists. If the universe could curve back on itself like a circle, there would be space left outside the circle.

Perhaps the most astonishing flaw in Einstein's conception of the universe was his idea that the objects existing in the universe could not be in motion. Surely

Einstein was aware of the movement of the stars, planets, clouds, and wind. There is nothing existing in the universe that is not in motion. Yet, Einstein tried to prove that the universe is motionless. He invented the idea of the "cosmological constant"—the idea that there exists in the universe an "antigravity force" that counterbalances gravity and creates a motionless universe—a universe that neither expands nor contracts.

Einstein's idea of a cosmological constant was a farce. Einstein, himself, considered the idea of a cosmological constant to be the biggest blunder he ever made in his life.

Einstein went from conceiving a motionless universe to seeing objects traveling at very high speed in the universe. He wanted a way to measure their speeds. He realized that he needed a "constant," a thing that does not change to measure things that do change. He established the constancy of the speed of light in the universe. He said that the speed of light, which travels at 186,000 miles (300,000 kilometers) per second, is a universal constant. He explained that no matter who is measuring the speed of light, or under what conditions, the answer is always the same. He held that the speed of light is the highest possible speed that can occur in the universe. It is the universal speed limit. Today we know that this is not true. Atoms, the tiny particles of which light is composed, travel "many" times faster than light. I do not argue against light being a constant, only that it is not the fastest thing in the universe. Let it be understood that "electron atoms" whirl around the nucleus of the thom several billion times in a millionth of a second. The number of vibratory movements made by "Cesium atoms" in one second is 9,192,531,770. Cesium atoms are used in atomic clocks, I believe that the "gravitational effects" of the atoms must travel throughout the universe at nearly the same speed as do the atoms, many times faster than light.

When we use the atomic clock to measure the motion of things or events, Einstein's theory of the relativity of motion, based on the observer, collapses. Time becomes absolute and so does the measurement of motion. Telling time by the atomic clock makes Einstein's theory of relativity of time for different people erroneous. The time indicated by the atomic clock will not differ from person to person.

Einstein's theory of the relativity of time and motion based on the observer, and that gravitational effects cannot travel faster than the speed of light, is not an adequate basis for understanding our "Atomic" universe.

Einstein never conceived of the universe as being made up of eternal and indestructible material atoms and kenon (empty space). He conceived of the universe as being made up of matter and held matter to be a form of energy. He never explained what energy is, where it comes from, and how it causes matter to change and move. Let it be understood that all matter is an aggregate of atoms, and that atoms are the only forces in the universe. Let it also be understood that the word energy (from the Greek *energeia*, meaning "activity") is the name for the activity of atoms—or matter. Energy is produced by the movements of the atoms. When atoms are rearranging themselves and causing change in matter, they are carrying on energy (activity).

Einstein claimed that matter can be changed into energy and energy back into matter. He was wrong! How can a thing be changed into its own activity and its own activity be changed into it. This is like saying that the ocean can be changed into its waves and its waves back into it.

Einstein even claimed that energy can be created sometimes by destroying matter. This is nonsense! Matter cannot be destroyed, but only broken up into constituent atoms. If matter could be destroyed, energy would cease to exist. We do not live in a universe that subtracts matter from itself to add energy. We live in a universe that cannot be changed by addition or subtraction; it remains quantitatively the same.

From all that I have explained, surely you can see that Einstein has given us a false conception of the universe.

Essay #12
Why Quantum Theory and Quantum Mechanics are Wrong

As *The American Heritage School Dictionary* explains, the quantum theory is: A theory of physics based on a complex use of mathematics and proposed to explain the interactions of matter and energy at the atomic and subatomic levels. Its essential features include the idea that energy occurs in quanta and that particles and waves have certain properties in common.[32]

A *quantum* is defined as "an indivisible unit of energy, as a photon." The plural of quantum is *quanta*.[33]

To me, the most glaring flaw in the quantum theory is its proposal to explain the interaction of matter and energy at *the subatomic level*. This is because there is no such thing as a subatomic level. Again, I will reiterate: *The atom is the smallest particle of matter that can exist—the ultimate and smallest division of matter.*

The quantum theory claims that matter is not made up of solid atoms, and that the atom itself can be divided into many constituent particles. In consequence of this false claim, quantum theorists say that atoms are not the building blocks of matter.

Quantum theorists mistake the atom for the thom and consequently make wrong and false statements about the atom. The statements that they make about the atom apply to the thom instead.

Quantum theorists say that if we were to take a strip of aluminum or a piece of copper (elements) and subdivide them into smaller and smaller pieces, we would eventually come to a tiny particle which, if further subdivided, would no longer show the properties of the element. They call the smallest particle of an element that has all the properties of the element an atom, but it is a thom, not an atom.

According to quantum theorists, the quantum theory is very important in science because it can, in principle, predict all the chemical and physical properties of a substance. In practice, however, scientists are not completely successful in making such predictions. As the American physicist Lillian Hoddeson explains, "The mathematical equations involved in the application of the quantum theory are so complex that, except for a few special cases, scientists do not yet know how to solve them. To obtain equations that are solvable, they make certain assumptions about the structure of a [thom] that are not really true. But these assumptions lead them to equations that are simple enough to be solved with known methods. The results calculated in this fashion are not really exact. They are only approximations to the truth. However, they are better than no information at all."[34]

Quantum theorists say that in the subthomic world, particles do not obey fixed rules. Their individual movements, while statistically predictable, are uncertain. This, they say, is because we cannot measure both the position and momentum of an subthomic particle at the same time. This idea is expressed by "the Uncertainty Principle" developed by the German physicist Werner Heisenberg (1901-1976) in 1927. This principle states that if the position of a subthomic particle is known, then a determination of its motion will be uncertain; when the motion of the subthomic particle is known, the determination of its position will be uncertain.

In my way of thinking, there is a strong possibility that Heisenberg's uncertainty principle might break down when applied to atoms that have formed themselves into a very tight aggregate and make only vibrative movements in constantly held positions. Scientists could probably determine the positions held by atoms in such aggregates and at the same time determine their motion, just as they determine the number of vibratory movements of the element Cesium used in atomic clocks. The number of vibrations that Cesium makes in one second is 9,192,631,770.

Let it be understood that the quantum theory itself has no applicability to atoms themselves, which are not composed of energy and therefore cannot release energy. They create energy by their motion. According to the quantum theory, energy cannot be created or destroyed, but it can be changed from one form to another. That is simply not true. All forms of "so called" energy—heat energy, light energy, sound energy, electrical energy, mechanical energy, etc.—are created, or produced, by the motions of atoms. They are destroyed when the specific kinds of interactions of atoms creating them reach their durational end.

It is important for quantum theorists to realize that gravity is not a form of energy as they think, but rather an inherent fundamental *dynamis* and property of atoms that enable them to produce energy by their movements.

I hold that energy is the activity of atoms directed toward doing work, not the ability to do work or to change the shape or the phase of matter. Without the dynamis, or power, of atoms to do, act, or produce a certain effect, it would be impossible for heat, light, sound, electricity, and wind effects to exist.

In discussing quantum theory, I always end up discussing "quantum mechanics," a mathematical theory in physics that starts with the assumption that energy is not infinitely divisible and deals with "supposed" atomic structure and phenomena by the methods of quantum theory. According to this theory, electrons only move in *orbits* at fixed levels around the nucleus of an atom. When an electron gains a quantum of *energy*, it jumps a level. When it falls back a level, it gives off a quantum of light energy.

When "quantum mechanists," followers of the quantum theory, speak of electrons orbiting the nucleus of an atom, they are usually thinking of the solar system model of an atom, proposed in 1913 by the Danish physicist Niels Bohr, or the electron cloud model, proposed in 1928. The solar system model shows electrons traveling in fixed orbits like planets around the sun, considered the nucleus. The electron cloud model indicates the regions within the atom where electrons may be found. It is said that electrons are most likely to be where a cloud is darkest. Both models are wrong. Electrons are much larger than atoms. Atoms are the smallest particles of matter that can exist—the ultimate and smallest division of matter. If matter, including electrons, is repeatedly cut up, the end result will be uncuttable pieces of matter, atoms. Electrons have a mass of 9.11×10^{-28} grams, while atoms have a mass even smaller than that of the smallest *quark*, the U quark, which has a mass of approximately 1×10^{-27} grams. This shows that atoms are at least 1,000 times as small as electrons and that it is impossible for electrons to be subatomic particles.

In addition to this, atoms are absolutely solid particles, with no place for anything to exist inside of them. They can neither emit nor absorb energy, whatever quantum mechanists consider that to be.

Quantum mechanists need to understand that quantum mechanics is not a description of the behavior of matter on an atomic scale, but rather on a "thomic scale." All subthomic particles—quarks, electrons, protons, and neutrons—are aggregates of atoms and that their behavior arise from the interaction of the atoms of which they are composed. They also need to understand that what they call energy—heat, light, sound, electricity, wind, etc.—is *effect* produced by the interaction of atoms. Energy is not a thing, only the activity of a thing. The only things that exist are atoms, which do not possess energy, but rather produce it by their movements. Energy is not something that can be separated from atoms and interact with something; only atoms can interact and cause change in matter (objects composed of atoms). All changes in matter can be explained in terms of the self-moving, eternal, material atoms shifting their arrangement and position within *kena* (empty spaces). With all of this understood, quantum mechanists will be able to give a better description of the behavior of matter in all its details and, in particular, of the happenings on the thomic scale. They will be able to calculate precisely the "atomic" properties of thoms, molecules, and materials, which will greatly increase their ability to design electronic components, new materials, and drugs.

I am aware that without the present quantum theory followed by quantum mechanists, there would be no computers, cellular telephones, or many other recent inventions. Nevertheless, I am also aware that the quantum theory is a false energy theory which only works superficially and can never penetrate deeper into the understanding of the atomic nature of matter, which is absolutely necessary to the growth or progress of understanding what everything is basically made of and how everything ultimately works.

Essay #13
Biologists' False Claim Against the Theory of Spontaneous Generation

Spontaneous generation is a theory that attempts to explain the origin of life from nonliving matter. Biologists today, except for a few, falsely claim that this theory has been long discredited and that it has been replaced with *biogenesis*, a theory that living things come only from other living things. It is understood that all present life on Earth is the result of the reproduction of pre-existing life. The theory of biogenesis is only superficially true. I say this because it does not explain the origin of life itself.

Let it be understood that biogenesis cannot replace the theory of spontaneous generation. Our Earth, as it was formed by the atoms of which it is composed, was certainly lifeless—yet life appeared. Consequently, life did indeed arise by means of spontaneous generation from nonliving matter, which is composed of nonliving atoms. The only difference between living things and nonliving things is the arrangement of their atoms. This has been explained by the Atomists since the ancient Greek Atomist Democritus (c.460-c.370 B.C.).

Democritus explained that randomly moving atoms came together to form the first living organisms on Earth, the ultimate origins of all the diverse life forms on Earth. All the different species of living things—including humans, animals, fish, insects, plants, and microorganisms—evolved from the first "atomic" forms of life. This process took over three billion years. Many of the stages in the development of living things can be traced in the fossils laid down in rock strata of different ages. Let me explain how all *present* life on Earth is the result of reproduction of pre-existing "atomic" life.

About 5 billion years ago, our solar system was a whirling mass of gas atoms and dust atoms. The sun and planets formed from this mass. Earth is thought to be about 4.6 billion years old. Rocks found in Australia, which are more than 3.5

billion years old, contain fossils of once-living organisms. The atmosphere of primeval Earth was made up of gases of atoms similar to ammonia, hydrogen, methane, and water vapor. Lightning and ultraviolet rays of atoms from the sun collided with these primeval gases of atoms to combine and form the chemical aggregates of atoms from which living things are made. Atoms aggregated to form *quarks,* quarks aggregated to form *protons, neutrons, electrons, neutrinos, etc.* Protons, neutrons, and electrons interacted to form *thoms,* thoms interacted to form *molecules,* molecules interacted and produced the "secret" chemical mixture of life, or magic life "stuff," known as *protoplasm.* Protoplasm is made of carbon atoms, hydrogen atoms, oxygen atoms, nitrogen atoms, phosphorous atoms, and several other atomic substances, or elements. They are joined together in compounds that may be simple, like water, or they may be very complex. Some of the compounds in protoplasm contain as many as two thousand thoms belonging to five or six different elements. No one has yet discovered how all these thoms are arranged in a single molecule. A single unit of protoplasm forms a *cell.* A cell is the basic atomic unit of structure and function for a living thing. Humans, animals, fish, insects, plants, and microorganisms are made up of cells. A group of similar cells working together to perform a specific job is called a *tissue.* Nerve cells are grouped together to form nerve tissue, which carries messages. The thick-walled cells in a tree trunk form a tissue that supports the tree. Tissues of various kinds are often grouped together to form an *organ.* Your heart, your eyes, your feet, and even your skin are all organs. Plants also have organs, leaves and roots are examples. Organs working together are often combined into *systems.* A system carries on a complete life function for the living thing. Your *circulatory system* carries blood through your body, delivering food and oxygen to all the body's cells. It also carries wastes away from the cells. The *digestive system* breaks down food into simpler parts so that body cells can use it. The *respiratory system* takes in oxygen and combines it with food to get power. The *nervous system* carries messages all over the body. All systems put together make up the human body—a complex living system.

The spontaneous generation of living systems from non-living atoms is carried perpetually onward by the eternally and constantly moving atoms of which all living systems, or things, are composed. The moving atoms that form *genes* cause genes to change and to produce change. Genes are the structures on *chromosomes* that determine genetic traits. Genes and chromosomes are found in the reproductive cells of plants and animals. These cells divide and go on from one generation to the next, carrying the genes that produce old or new heredity traits. Changes produced by genes, therefore, are passed on. This is a continuous process that causes *a slow, gradual development or change in appearance of living things over time.* Such a change is what scientists call "biological evolution." Fossils, or the remains or traces of things that lived ages ago, in ancient strata show that living things have been changing for at least fifteen hundred million years.

As the American philosopher Reuben Abel explains, "Modern biochemistry has established that all genes of all living creatures are made up of the same substances (DNA, RNA, and proteins.) A man's genes differ from a dog's, say only in the way they are arranged. The same contractile protein produces the streaming motion of the amoeba and the moving finger muscles of the pianist. Heredity operates in the same way in plants, bacteria, and human beings. This chemical unity of all life makes it conceivable that life originated only once![35]

For me, the combined facts of the Atomic theory of spontaneous generation, fossil revealments, and the chemical unity of life, provide substantial proof for biological evolution. I believe that all animal species are related to one another. I accept the biological evolutionary theory that says humans come from the same group of primates as the chimpanzees and other apes.

To conclude this essay, I will give this final explanation: I believe in the spontaneous generation of life from nonliving matter because I understand that all living matter is made up of nonliving atoms. I also understand that, in reality, nothing exists but atoms and kenon (empty space) and that the interaction of atoms ultimately causes all the different phenomena in nature, including the rise of living matter.

Atoms are self-moving, and they move randomly. This proves that there is nothing controlling their interaction, and that all of their interactions, and the effects produced by their interactions are spontaneous.

Essay #14
Myths About Black Holes and Baby Universes

According to the English theoretical physicist Stephen Hawking, the concept of what we now call a black hole goes back more than two hundred years, but the name *black hole* was introduced by the American physicist John Wheeler in 1967. In space we find that there are "black holes"—very dense areas of matter with tremendous gravitational power. The gravity in a black hole is so powerful that nothing, not even light, can escape from it. This makes the black hole invisible. Star-size black holes are believed to be the last stages of giant stars that have collapsed into themselves.

In astronomy, the study of the moon, stars, and other objects in space, we are told that: The most massive stars—those having more than 40 times the mass of the sun—become *black holes* when they die. After this kind of star becomes a supernova, more than five times the mass of the sun may be left. The gravity of this mass is so strong that gas is pulled inward, packing the gas into smaller and smaller space. Eventually five times as much as the sun becomes packed within the diameter. At that point, the gravity is so strong that nothing can escape, not even light. The remains of the star becomes a black hole.[36]

From 1970 to 1974, Hawking investigated black holes and in 1974 made a surprising discovery: black holes are not completely black and that a black hole emits radiation as if it were a hot body. He explained that particles that enter the gravitational field of a black hole at the speed of light could pass through the hole. This is because black holes are small and particles that enter them at the speed of light only have to go a short distance to get through. Hawking explained that: "As a black hole gives off particles and radiation, it will lose mass. This will cause the black hole to get smaller and send out particles more rapidly. Eventually, it will get down to zero mass and will disappear completely. What will happen then to the objects ... that have fallen into the black hole? According to some recent work of mine, the answer is that they will go off into a little baby universe

51

of their own. A small, self-contained universe branches off from our region of the universe. This baby universe may join on again to our region of spacetime. If it does, it would appear to us to be another black hole that formed and then evaporated. Particles that fell into one black hole would appear as particles emitted by the other black hole, and vice versa." [37]

Hawking emphasized that what comes out of a black hole will be different from what fell in. He said that only energy will be the same.

Hawking's talk about "baby universes" is *mythical*. The particles that come out of a black hole do not go off into baby universes. There is only one universe, which Atomists call "the all." Let it be understood that no region of the universe is a baby universe and that black holes are not pathways to baby universes. There is no such thing as the infinite universe being made up of finite universes. The very definition of the word *universe*, as meaning "the all," makes that impossible. Let it also be understood that the particles that come out of a black hole do not come out as energy. Particles are not composed of energy. They do not possess energy from their sheer existence. The word *energy* (from Greek *energeia*, meaning "activity") is the Atomists' name for the activity, or interaction, of atoms. Atoms are the smallest particles of matter that can exist—the ultimate and smallest division of matter. The interaction of atoms ultimately causes all the different phenomena in nature, including the occurrence of energy.

The universe is composed of atoms and kenon (empty space); there is nothing else. The matter that falls into a black hole is an aggregate of atoms. Black holes are nothing but nature's own "atomizers"—gravitational fields that break up or separate matter into its constituent atoms. All matter pulled into a black hole is probably atomized and eventually forced out into the kenon (void) beyond the black hole. Since atoms are too small to be seen, this event is not detectable. The event is gradual and continuous.

Only by atomizing the matter that it pulls into its hole, and by releasing the atoms, can black holes continue to pull in the huge amount of matter surrounding them. The light pulled into black holes is not trapped, it is atomized. All light is an aggregate of atoms. When the black holes atomize the light that falls into them, the light disappears.

Now that you know that black holes are atomizers of matter, you have more understanding of how our atomic universe works.

Concluding Thoughts

I have introduced you to Atomism, also known as the Atomic theory, as a way to put an end to the teaching of pseudo-science in our schools, colleges, and universities, and as a way to acquaint the scientific community with the proper way to do physics. Physics, which is the natural science that deals with the study of what things are made of and how they change, constitutes the basic or starting point of scientific knowledge. All other natural sciences—biology, chemistry, geology, etc.—depend on physics for the foundations of their knowledge.

I am asking you to believe in the existence of the atom as described in this book and the following basic principles of Atomism:

- Nothing can come into existence out of nothing. Therefore, if something comes into existence, it must come from something previously existing.

- Nothing can be destroyed into nothing, but only broken up into constituent atoms.

- Everything that exists is made up of atoms and kenon.

- Nothing can come into existence without atoms.

- Atoms and kenon are the only ultimate realities.

- All other things are properties or accidents of atoms and kenon.

- Kenon is the pure empty space in which or at which there is absolutely nothing present.

- Kenon possesses no dynameis (powers), no potentiality, and no tangibility.

- Atoms are absolutely solid, everlasting, and simple.

- Being absolutely solid, atoms are indivisible.

- Atoms cannot undergo change.

- Although physically indivisible, atoms have parts that are the *minima* of extension and magnitude.

If you believe in the existence of the atoms as described in this book and the basic principles of Atomism, you have become an Atomist. You no longer accept the splittable atom of modern physics as an atom. You also no longer accept the atomic theory of modern physics as the true atomic theory.

Whether you are a member of the scientific community or of the general public, I am sure that you realize that Atomism is an irrefutable theory of reality and causality regarding nature, and that it has great value to anyone who is seriously interested in knowing about the nature of the universe and wants final and realistic answers to the questions: "Of what does reality consist?" "What is the universe made of?" "How does the universe work, or operate?" As Atomism explains, reality consist of atoms and the kenon, the universe is made up of atoms and kenon, and the universe operates atomically.

Recall that Atomism explains that the interaction of atoms ultimately causes all of the different phenomena in nature. Also recall that all changes in matter, the ultimate substance that makes up the material universe, are the result of self-moving, eternal, and material atoms moving and combining in different ways. I am sure that you understand that since every change in matter has an atomic cause, atomic explanations of changes in matter are the only correct explanations.

With this understanding, you can now see that the theory of Atomism supersedes energy theory of matter. Remember that energy theory claims that matter and energy are equivalent, and that matter can be changed into energy and energy can be changed into matter. You can now explain to energy theorists that it is impossible for matter and energy to be equivalent because matter has mass and occupies space and energy does not and cannot. You can explain that matter can never change to something that has no mass and that anything with mass is matter. Matter can only change from one form of matter into another. It can never be changed into something that is non-material, but only broken up into constituent atoms. Nothing which is non-material can exist, except kenon.

You have learned from the theory of Atomism that energy can only be the "activity" of matter, or atoms, and that it is wrong to think that matter can be destroyed to create its own activity. You now understand that energy, being the activity, or activity of matter, cannot act—or cause anything to happen. It cannot do work such as moving things or giving heat or light. It cannot change the shape or phase of matter. It cannot make living things grow. You can explain to energy theorists that it is atoms which act—and cause things to happen. It is senseless to talk about what energy can do, and bypass the atoms altogether. We cannot dispense with atoms and talk about their energeia or energy.

You can now explain to the energy theorists that energy can be created and destroyed by the movements of atoms, and that the energy created by the movements of atoms vary in duration. You can also explain that energy cannot be changed from one form to another, it is the atoms that change their energy into different energy.

We say, let no one think that scientists can bypass the atoms, or dispense with them, and explain how matter changes form and moves from place to place. We can explain to scientists that all changes in matter are the result of atoms moving and combining in different ways in kenon. We can explain that motion is a fundamental property of atoms, and that all changes in nature are reducible to the movements of atoms in the kenon.

We can explain to scientists that when matter seems to disappear, all that happens is that it breaks up into its tiny imperceptible atoms. When the mass of matter increases, it is because matter has absorbed atoms. When the mass of matter decreases, it is because matter has loss some of its atoms, but the atoms themselves have not been destroyed.

We must request that all scientists acknowledge the fact that matter is made up of eternal and indestructible atoms, and that the mass of an object is a measure of the number and shapes of atoms that make it up.

We must explain to scientists that atoms have dynamis from their sheer existence, but not energy. It is a mistake to think that it is the energy of atoms that has the ability to do work; it is the dynamis of atoms that gives them this ability. Modern physicists confuse energy with dynamis.

We must explain to physicists that since everything is made up of atoms and kenon, and since nothing exists but atoms and kenon, and since the interaction of atoms within the kenon ultimately causes all the different phenomenon in nature, the proper way to do physics is to study the interaction of atoms. We must try to convince them that everything can be understood by describing the size, shape, solidity, *varos* (weight, or force of mass), gravity, position, arrangement, repellency, movement, and dynamis of the atoms.

We must explain to physicists that doing physics by studying the interaction of atoms has great value and benefits. It will help them give an account of reality and make physics more than a set of formulas that predict what we see in an experiment, as it should be. It will help them explain that cause and effect is a universal law of atomic reality. It will help them explain that all actions are caused by atoms, and that the nature of an action is caused and determined by the nature of the atoms that act, and that a thing cannot act in contradiction to its atomic nature. It will help them understand, explain, and predict the behavior of matter.

This will allow them to explain why different materials have different characteristics, predict how a material will change when heated or cooled, and figure out how to combine thoms and molecules to make new and useful materials. It will help them understand how mass tends to be conserved when matter changes form. This will help them describe changes that they observe in the natural world and explain where matter goes when it seems to disappear. It will help them understand that matter changes form but is never destroyed. It will help them predict all the chemical and physical properties of a substance. This will help them to predict what will happen when matter changes form, direct the energeia of matter in useful ways, and build and improve machines. It will help them understand how each of the atomic forces at work in nature is necessary for us to do the things that we do. It will help them understand the machinery of gravity and that atoms are the only forces in the universe. It will help them understand that a property of an atom cannot be detached or separated from an atom. This will allow them to predict how objects will move, design machines that perform complex tasks, and predict where planets and stars will be in the sky from one night to the next. It will help them understand biological evolution. This will allow them to explain how atoms combine to form thoms, how thoms combine to form molecules, how molecules combine to form cells, how cells combine to form tissue, how tissue combines to form organs, and how organs combine to form living systems. It will help them understand that "thought" is a movement of atoms in the body. This will allow them to figure out how to make atoms arrange themselves in the body to produce thoughts. It will help them understand how the tiniest of things work. This will enable them to construct "atomic micro and nanotechnology". It will give them the final answers to the questions: "What is the universe and all things in it made of?" "How does the universe and all things in it work?"

I will now end my concluding thoughts about the theory of Atomism and the important benefits of properly doing physics by studying the interaction of atoms, which ultimately causes all of the different phenomena in nature. If physicists choose to continue doing physics by studying the "supposed" interaction of matter and energy, and thereby produce pseudo-science, I leave it to them to experience the inevitable deterioration of their physics.

I believe that as more and more physicists become well acquainted with the theory of Atomism, its basic principles will become the foundational tenets of modern physics, and that many people in the general public will become interested in studying physics. I believe that you, yourself, will be one of them.

References and Notes

Front cover quote: Quoted in Milton Dank, *Albert Einstein* (Franklin Watts, New York, 1983), p.80.

Introduction

[1] Democritus, *Diogenes Laertius*, lives IX, 44-5

[2] Richard P.Feyman, *Six Easy Pieces* (Cambridge, Massachusetts: Perseus Books, 1995), p. 71

[3] This is because the universe is made up of atoms and operates atomically.

[4] The Atomists' argument for the existence of atoms is based on this definition of the word "atom." They argue that all matter can be divided into tiny particles that cannot be further divided, atoms. In modern chemistry, the word "atom" is

erties of the element.

[5] The most literal word in English for *kenon* is "void."

[6] The most literal word in English for *dynamis* is "power."

[7] "Varos" does not mean the force of gravity on an object but only the downward force of an object's mass on empty space or on another object.

Essay #1

[8] *Philosophical Essays* (1910) No. 4

Essay #2

[9] Quoted in Roger B. Beck, Linda Black, et al., *World History: Patterns of Interaction* (Evanston, IL: McDougal Littell, 1999), p. 125.

[10] Peter A. Angles, *The Harper Collins dictionary of Philosophy*, 2nd Edition (New York: Harper Collins Publishers, 1992), p. 24.

[11] Lucretius, *On the Nature of the Universe*, translated by R.E. Latham and revised with a New Introduction and Notes by John Godwin (London: Penguin books Ltd., 1951), p. x.

[12] Lucretius, *On the Nature of Things*, translated, with Introduction and Notes, by Martin Ferguson Smith (Indianapolis: Hackett Publishing Company, Inc., 2001) p. xxvi.

[13] Angles, *The Harper Collins dictionary of Philosophy*, 1992, pp. 333-34

[14] Lucretius, *On the Nature of the Universe*, 1951, pp. 18-19

[15] Ibid, p. 23

[16] Quoted by G.E.R. Lloyd: "Leucippus and Democritus," *The Encyclopedia of Philosophy* (new York: the Macmillian company and Free Press, 1967), Volume four, pp. 447-48

[17] Bryan Magee, *The Story of Philosophy* (London: Dorling Kindersley Limited, 2001), p.44

[18] *Masterpieces of World Literature*, edited by Frank N. Magill (New York, NY: Harper Collins Publishers, 1989), p. 65

[19] Lucretius, *On the Nature of Things*, 2001, p. xxvi

[20] Anthony Gottlieb, *The Dream of Reason* (New York: W.W. Norton & company, 2000), p. 95

[21] The solar system model of the thom leaves it without a surface. This makes the existence of the thom problematic.

[22] Gottlieb, *The Dream of Reason*, p. 104.

Essay #3

[23] *Random House Webster's Unabridged Dictionary* (New York: Random House, Inc., 1998), p. 1884.

[24] *Oxford Dictionary of Physics* (New York: Oxford University Press Inc., 2003), p. 476.

Essay #5

[25] Feynman, *Six Easy Pieces*, 1995, p.71

[26] *Dynamis* should be used in modern physics as a corrective for the idea of energy.

Essay #6

[27] Understanding the four causes of the motion of atoms is necessary for understanding how the world works.

Essay #7

[28] http://www.quotatiionspage.com/subjects/truth/ArthurSchopenhauer

[29] Feynman, *Six Easy Pieces*, 1995, p.71
[30] Angeles, *The Harper Collins Dictionary of Philosophy*, 2nd edition, p. 77
[31] Ibid.

Essay #12

[32] *The American Heritage Students Dictionary* (Boston: Houghton Mifflin company, 1986), p. 720
[33] Ibid.
[34] Lillian Hoddeson, "Quantum Theory," *The New Book of Popular Science*, Vol. 3 (Danbury Connecticut: Grolier, 2004), p. 322

Essay #13

[35] Reuben Abel, *MAN is the Measure* (New York: The Free Press, 1976), p. 148.

Essay #14

[36] "Astronomy," *Science Explorer* (Needham, Massachusetts: Prentice Hall, Inc., 2002), p. 115
[37] Stephen Hawking, *Black Holes ad Baby Universes and Other Essays* (New York: Bantam Books, 1993), p. 121.

About the Author

Mohammed Abu-Bakr is a retired teacher from George Washington High School, a Denver public school, for which he was a student advisor and special education teacher. He taught physical and biological sciences along with other core courses. He was also a member of the Extended Studies Staff of Adams State College located in Alamosa, Colorado, for which he taught education and philosophy courses. He is an Atomist—a believer in the classical natural philosophy of Atomism, also known as "the Atomic Theory," which originated the idea of atoms.

Index

Aristotle, 6, 11, 31
Atom
    cesium atoms, 42
    characteristics of, 7
    gravity and, 39-40
    and kenon, xiv, 3–4, 17, 50, 52–55
    meaning and definition of, xiv, 7
    momentum of, 32–33
    motion, causes of, xv, 28–29
    origin of life and, 48–49
    properties of, xv, 2, 7
    in quantum theory, 44–46
    size of, 2, 28, 36
    Thomson's model of, 14–15
Atomic theory
    atomists and, 53–56
    Boscovich's view on, 13–14
    modern physicists and, 18
    in science textbooks, xvii, 10
    spontaneous generation of life and,
        50
Atomism
    development of, 6–19
    during Middle Ages, 11–12
    key words of, xvi
    meaning of, xiv, 1
    principles of, 3–4, 53
    as theory of reality and causality, 54
Atomists
    concept of atoms by, xiv, 7–11, 14
    motion of atom and, 28
    notion of void and, 8–9
    quantum theory and, 44–47
    view on gravitation, 39–40
    view on light, 34–36
    view on origin of life, 48

    view on time, 37–38
    view on universe, 41–43
Atoms and kenon
    as constituents of the universe, xii–xv,
        1, 17, 52, 54
    as ultimate reality, 3, 53

Baby universes
    and black hole, 51–52
    Hawking's view on, 52
Big Bang theory
    collapse of universe, 23
    existence of matter and energy, 20
    formation of atom, 21
    refutation of, 20–23
Black holes
    and baby universe, 52
    concept of, 51
    myths about, 52
Bohr, Niels, 15–16, 46
Boscovich, Rudjero Giuseppe, 12–13

Chadwick, James, 16
Curvature of space
    Einstein's view, refuting, 39–40
    and gravity, 39–40

Dalton, John, 13–15
Democritus, 1, 6–8, 10, 12–13, 16, 48
Dynamis
    atom's motion and, 28–29
    calculating, 31
    energy theorists and, xvi
    Greek origin of, 31
    meaning of, xv
    as property of atom, 3, 7

www.ingramcontent.com/pod-product-compliance
Lightning Source LLC
Chambersburg PA
CBHW021004180526
45163CB00005B/1888